Loudspeaker
Handbook

JOIN US ON THE INTERNET
WWW: http://www.thomson.com
EMAIL: findit@kiosk.thomson.com

thomson.com is the on-line portal for the products, services and resources available from International Thomson Publishing (ITP). This Internet kiosk gives users immediate access to more than 34 ITP publishers and over 20,000 products. Through *thomson.com* Internet users can search catalogs, examine subject-specific resource centers and subscribe to electronic discussion lists. You can purchase ITP products from your local bookseller, or directly through *thomson.com.*

Visit Chapman & Hall's Internet Resource Center for information on our new publications, links to useful sites on the World Wide Web and an opportunity to join our e-mail mailing list. Point your browser to: **http://www.chaphall.com** or **http://www.chaphall.com/chaphall/electeng.html** for Electrical Engineering

A service of

Loudspeaker Handbook

John M. Eargle

CHAPMAN & HALL

I(T)P® International Thomson Publishing

New York • Albany • Bonn • Boston • Cincinnati • Detroit • London • Madrid • Melbourne
Mexico City • Pacific Grove • Paris • San Francisco • Singapore • Tokyo • Toronto • Washington

Cover design: Saïd Sayrafiezadeh, emDASH inc.
Art direction: Andrea Meyer

Printed in the United States of America

Chapman & Hall
115 Fifth Avenue
New York, NY 10003

Chapman & Hall
2-6 Boundary Row
London SE1 8HN
England

Thomas Nelson Australia
102 Dodds Street
South Melbourne, 3205
Victoria, Australia

Chapman & Hall GmbH
Postfach 100 263
D-69442 Weinheim
Germany

International Thomson Editores
Campos Eliseos 385, Piso 7
Col. Polanco
11560 Mexico D.F
Mexico

International Thomson Publishing–Japan
Hirakawacho-cho Kyowa Building, 3F
1-2-1 Hirakawacho-cho
Chiyoda-ku, 102 Tokyo
Japan

International Thomson Publishing Asia
221 Henderson Road #05-10
Henderson Building
Singapore 0315

1 2 3 4 5 6 7 8 9 10 XXX 01 00 99 98 97

Library of Congress Cataloging-in-Publication Data

Eargle, John.
    Loudspeaker handbook / John M. Eargle.
        p.   cm.
    Includes bibliographical references and index.
    ISBN 0-412-09721-4
        1. Louspeakers—Handbooks, manuals, etc.   I. Title.
    TK5983.E27   1996
        621.382'84—dc20   96-22852
                                                                 CIP

British Library Cataloguing in Publication Data available

To order this or any other Chapman & Hall book, please contact **International Thomson Publishing, 7625 Empire Drive, Florence, KY 41042.** Phone: (606) 525-6600 or 1-800-842-3636. Fax: (606) 525-7778. e-mail: order@chaphall.com.

For a complete listing of Chapman & Hall titles, send your request to **Chapman & Hall, Dept. BC, 115 Fifth Avenue, New York, NY 10003.**

Contents

Preface

The prospect of writing a book on loudspeakers is a daunting one, since only a multivolume encyclopedia could truly do justice to the subject. Authors writing about this subject have generally concentrated on their own areas of expertise, often covering their own specific topics in great detail. This book is no exception; the author's background is largely in professional loudspeaker application and specification, and the emphasis in this book is on basic component design, operation, measurement, and system concepts.

The book falls largely into two sections; the first (Chapters 1–9) emphasizing the building blocks of the art and the second (Chapters 10–16) emphasizing applications, measurements, and modeling. While a thorough understanding of the book requires a basic knowledge of complex algebra, much of it is understandable through referring to the graphics. Every attempt has been made to keep graphics clear and intuitive.

Chapter 1 deals with the basic electro-mechano-acoustical chain between input to the loudspeaker and its useful output, with emphasis on the governing equations and equivalent circuits.

Chapter 2 is a survey of cone and dome drivers, the stock-in-trade of the industry. They are discussed in terms of type, design, performance, and performance limits.

Chapter 3 deals with magnetics. Once a source of difficulty in loudspeaker design, magnetics today yields easily to modeling techniques.

Chapter 4 discusses low-frequency (LF) system performance, primarily from the viewpoint of Thiele-Small parameters. We also discuss some of the multi-chamber LF systems that became popular during the eighties.

Chapter 5 is devoted to systems concepts. There is more diversity in this area than in any other aspect of loudspeaker design. In this chapter we discuss basic network types, baffle layout considerations, component matching, and system power response.

Chapter 6 discusses line and planar loudspeaker arrays. Such loudspeaker types radiate differently than individual cones and domes. We study arrays of dynamic drivers, as well as electrostatic and magnetic panels.

Chapter 7 covers horns and their drivers in detail. Horn systems are essential to professional sound applications and have formed a key element of electroacoustics since its inception.

Chapter 8 discusses the electronic interface, including the proper matching of amplifiers to drivers, series and parallel operation of amplifiers, aspects of multiamplification, and line losses.

Chapter 9 covers the performance shifts in loudspeakers due to the heating that occurs in high-level applications. Techniques for minimizing these effects are examined.

Chapter 10, 11, and 12 examine the application areas of recording and broadcast monitoring, sound reinforcement, and motion picture/video, respectively. Current practice in these areas is reviewed in detail.

Chapter 13 is devoted to loudspeaker measurements and modeling. A quiet revolution has taken place in recent years as measurement systems based on digital transform techniques have virtually displaced analog systems. Both methods are discussed, along with various techniques for modeling loudspeaker performance.

Chapter 14 discusses loudspeaker specifications for professional applications. Data presentation is an important first step in this area. We review current professional standards.

Chapter 15 discusses the home listening environment. Knowing where—and where not—to locate loudspeakers in the home often makes the difference between mediocre and excellent performance. We discuss some useful techniques here.

Chapter 16 discusses exotic transducers. These devices have always been on the fringes of the loudspeaker business, and some have been highly successful. We discuss many types in this chapter.

The author would like to thank his many colleagues at JBL and other Harman International companies for their direct help in gathering material for this book and for stimulating discussions over the years. Further recognition is given to those manufacturers who have provided illustrative materials for the book. They are cited in the figure captions.

John Eargle,
April 1996

Electroacoustical Engineering Fundamentals

1.1 Introduction

In this chapter we will develop the basic electro-mechano-acoustical model of a simple cone loudspeaker, detailing how the device attains its frequency region of flat power response and reference efficiency. We will also discuss the basic directional properties of cone loudspeakers under varying conditions of baffling. A number of assumptions will be made based on the physics of the acoustic wave equation, drawing on primary references in the literature as needed. In the way of terminology, a loudspeaker mechanism is generically a transducer, a device that changes power or energy in one form to another. More commonly, loudspeaker mechanisms are referred to as *drivers*, while the term *loudspeaker* is generally reserved for the complete system.

The dynamic cone driver is based on work originally described by Siemens (1874). Perhaps the seminal paper on the modern form the cone driver has taken is that of Rice and Kellogg (1925), in which the authors describe the specific roles of cone resonance and radiation resistance in attaining a wide frequency range of uniform power response. Over the nearly three quarters of a century since that time, the cone driver has seen countless variations and improvements, and yet remains clearly what it was at the outset. Needless to say, the cone driver has occupied a central position in a vast complex of consumer inventions and applications that continues today.

1.2 A Simple Electrical Series Resonant Circuit

Figure 1-1a shows a series electrical circuit with three passive elements and an active, time varying voltage generator, *e(t)*. The equation relating the impedance of the circuit, the current, *i(t)*, flowing through it, and the applied voltage is:

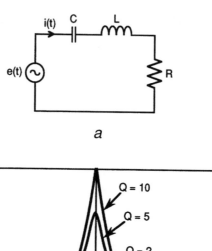

a

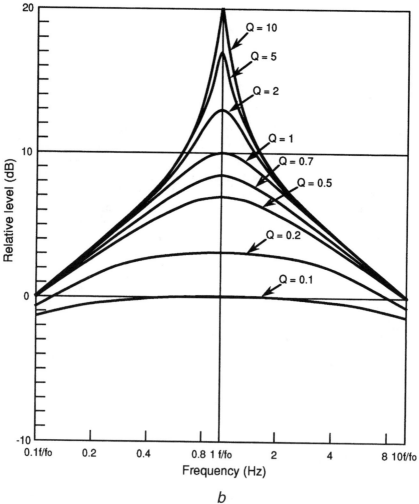

b

Figure 1-1. Example of electrical resonance. A series LCR (inductance, capacitance, resistance) circuit (*a*); family of resonance curves for different values of Q (*b*).

$$e(t) = i(t)(1/j\omega C + j\omega L + R) \tag{1.1}$$

where:
$\omega = 2\pi f$ (Hz)
j = square root of -1
C = capacitance (F)
L = inductance (H)
R = resistance (Ω)

We can also define the resonance frequency, f_0, of the circuit:

$$f_0 = 1/2\pi\sqrt{1/LC} \tag{1.2}$$

This is the driving frequency at which the reactive terms, $1/j\omega C$ and $j\omega L$, are equal and cancel each other, leaving only the resistive term in the right half of Equation (1.1). At this frequency the current through the circuit will be maximum.

The capacitive and inductive elements are reactive; they store power but do not consume it; only the resistive element dissipates power, and that is given by:

$$W = \frac{[e(t)]^2}{R} = [i(t)]^2 \tag{1.3}$$

If we plot the current flowing through the circuit as we vary the driving frequency of $e(t)$, we will get a curve resembling one of those shown in Figure 1-1b. The value of Q, the sharpness of the resonance curve, is given by:

$$Q = \omega_0 L/R \tag{1.4}$$

$\omega_0 L$ is the value of inductive reactance at the resonance frequency. If R is small relative to $\omega_0 L$ then the Q will be high, with a characteristic peak in the shape of the curve. If R is large, the shape of the curve will be smoother, as indicated by the lower values of Q in the family of curves.

We have shown electrical resonance in this example, and we will now move on to an example of mechanical resonance.

1.3 A Simple Mechanical Resonant System

Figure 1-2a shows a mechanical arrangement in which a rigid platform, free to move vertically, is suspended on spring and damping (resistive) elements. The mass of the platform is shown as a lumped value, M, and the platform is constrained to being driven up and down by a force generator, $f(t)$.

If we vary the driving frequency of the force generator we will soon discover that the velocity, $u(t)$, of the platform will be maximum at some resonance

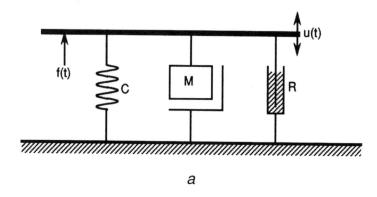

a

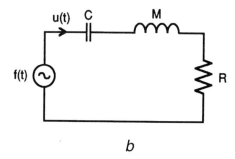

b

Figure 1-2.   A mechanical resonant circuit (a); impedance analogy of mechanical circuit (b).

frequency, f_0, and we will observe a family of resonance curves the same as those shown in Figure 1-1*b*.

The equation relating driving force, platform velocity, and mechanical impedance is:

$$f(t) = u(t)(1/j\omega C + j\omega M + R) \tag{1.5}$$

where:

C = mechanical compliance (m/N)
M = mechanical mass (kg)
R = mechanical resistance (damping), [N-s/m (mechanical ohms)].

Both electrical and mechanical circuits are described by equations of the same form and are said to be equivalent. Mass is analogous to inductance, compliance to capacitance, and damping to resistance. Force is analogous to voltage, and velocity to current.

Power dissipated in the damping element is:

$$W = \frac{[f(t)]^2}{R} \tag{1.6}$$

1.4 Impedance and Mobility Analogies

The mechanical analogy discussed here is called the *impedance analogy*. It is easy to understand because the related quantities are intuitively obvious:

Mechanical	Electrical
Force	Voltage
Velocity	Current
Mass	Inductance
Compliance	Capacitance
Damping	Resistance

However, from the viewpoint of constructing equivalent circuits, the so-called *mobility analogy* may be more useful to implement. Note the example in Figure 1-3. Here, the actual mechanical circuit we dealt with in Figure 1-2 is paired with its mobility analogy. The two are virtually the same in form, and herein lies the usefulness of the mobility analogy; the equivalent circuit can be drawn directly by inspection.

In order to do this, the roles have been switched, as given below:

Mechanical	Electrical
Velocity	Voltage
Force	Current
Compliance	Inductance
Mass	Capacitance
Responsiveness (r)	Resistance (R)

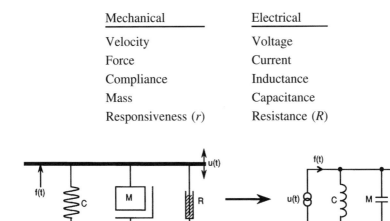

Figure 1-3. Illustration of a mobility analogy of the mechanical circuit.

The mobility equivalent circuit is known as the *dual* of the impedance equivalent circuit, and the governing equations are of the same form.

Those readers who wish to explore equivalent circuits further are referred to Beranek (1954) and Olson (1957) for their excellent discussions of these analytic techniques.

1.5 Combining Electrical and Mechanical Domains

The laws of elementary magnetism provide a simple connection between the electrical and mechanical domains via the following equations:

$$f = Bli \qquad\qquad (1.7)$$
$$e = Blu \qquad\qquad (1.8)$$

where:
  B = magnetic field strength (T)
  l = length of conductor in magnetic field (m)
  f = force (N)
  u = velocity (m/s)
  i = current (A)
  e = voltage (V)

These relationships are shown in Figure 1-4 *a* and *b*. Together they form the basis for a simple transducer that allows us to convert voltage and current into mechanical force and velocity, as symbolized by the transformer shown at *c*.

Any moving-coil transducer is an example of this, from a tiny earphone to a large shaker table. The product of voltage and current is electrical power, and the product of force and velocity is mechanical power, both the same quantities in terms of their units.

We can now construct an electrical driving system for our prototype mechanical circuit, and this is shown in Figure 1-5*a*. The equivalent electrical circuit is shown in Figure 1-5*b*. This transducer is capable of transforming electrical power into mechanical power. It can be used for shaking things, or even for canceling vibrations. But as yet it cannot make sound.

1.6 Combining Mechanical and Acoustical Domains

In order to produce sound, a vibrating object or surface must have sufficient expanse or area. The size of a piano sounding board, the size of a bass viol, or the bell of a tuba are all examples of this. For very low frequencies, a vibrating surface may do little more than move air back and forth. At these low frequencies the surface and its associated air load present essentially a mass reactance to the

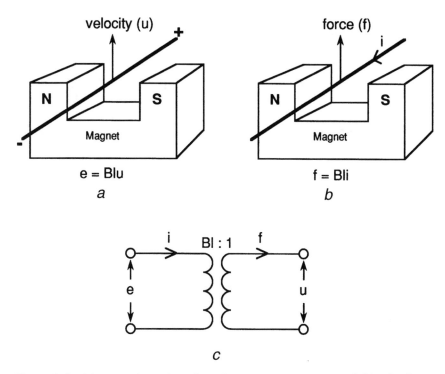

Figure 1-4. Magnetic relationships. A conductor moving in a magnetic field with velocity *u* will generate a voltage *e* across the conductor (*a*); a force *f* applied to a conductor in a magnetic field will produce a current *i* in the conductor (*b*); a transformer representation between electrical and mechanical domains that demonstrates *e* = *Blu* and *f* = *Bli* (*c*).

moving system, and little power can be radiated. The air mass associated with the cone is given by:

$$M_{\mathrm{air}} = 8a^3\rho_0/3 \tag{1.9}$$

where *a* is the radius of the piston (m) and ρ_0 is the density of air (1.18 kg/m³).

As the frequency of motion increases, sound begins to be radiated, as determined by radiation impedance. The nature of radiation impedance is fairly complex, and our discussion of it in this book is necessarily limited. Basically, it consists of two terms, the reactance of the air mass adjacent to the surface and the resistive term associated with the radiation of sound from the surface.

The resistive (R) and reactive (X) components of radiation impedance (Z) for a circular piston in a large baffle are shown in Figure 1-6, where $\rho_0 c$ is equal to 415 mechanical ohms (N-s/m), *a* is the radius of the piston (m), and *k* is equal to $2\pi/\lambda$, where λ is the radiated wavelength (m). The useful part of the impedance is of course the radiation resistance portion.

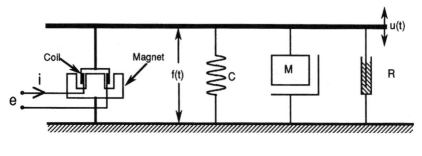

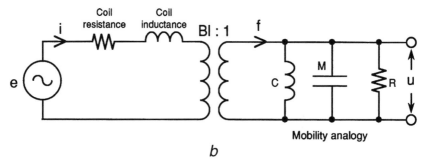

Figure 1-5. A simple electromechanical transducer. Mechanical circuit (*a*); electrical equivalent circuit using mobility analogy on the mechanical side (*b*).

We may think of *ka* as wavelength divided by the circumference of the piston. Note that the radiation resistance falls off below *ka* = 2 at a rate proportional to the square of frequency. This is equivalent to a low frequency (LF) rolloff of 12 dB per octave for constant displacement of the piston, or 6 dB per octave for constant velocity.

As we can see in Figure 1-6, the radiation resistance, R_A, in the region below *ka* = 1, is proportional to the square of frequency; that is, doubling the frequency in that range will produce a fourfold increase in R_A. We can say that:

$$R_A \propto f^2 \tag{1.10}$$

where the symbol $\propto$ indicates proportionality.

Recall from Equation (1.6) that

$$\text{Power} = [u(t)]^2 R$$

From Figure 1-7 we observe the mechanical resonance curve and note that in the mass controlled region above f_0:

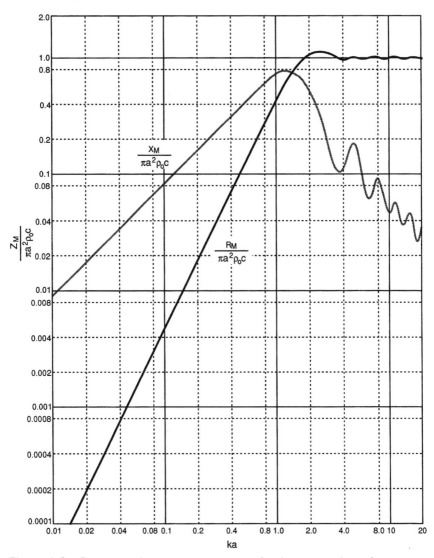

Figure 1-6. Resistive and reactive components of radiation impedance for a piston mounted in a large baffle. Ripples in response above $ka = 2$ are due to interference effects when wavelengths are small relative to the piston diameter.

$$u(t) \propto 1/f \tag{1.11}$$

and therefore:

$$[u(t)]^2 \propto 1/f^2 \tag{1.12}$$

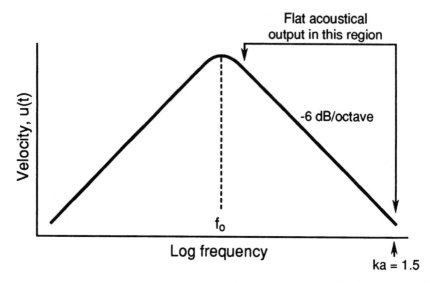

Figure 1-7.   Resonance curve of mechano-acoustical moving system. Uniform acoustical output can be obtained above f_0.

Since:

$$W_A \propto R_A[u(t)]^2$$

we can substitute terms and get

$$\text{Power} \propto f^2 \times 1/f^2 = 1 \text{ (constant)} \tag{1.13}$$

Thus, the radiated power from the cone will be constant over the frequency region between f_0 and a *ka* value of about 1.5–2.

The final electro-mechano-acoustical system, now drawn in the mechanical form of a loudspeaker, is shown as in Figure 1-8a. Note that we have now lumped the system moving mass into the cone, with elements of compliance (C_{MS}) and damping (R_{MS}) now associated with the system suspension. The cone mass (plus its associated air mass) is indicated as M_{MS} and the cone area as S_D. (The subscript MS indicates "moving system.")

The radiation impedance terms are shown as M_A and R_A. They are related to the mechanical circuit through the transformation ratio of 1:S, where S is the area of the cone.

The response is as shown at *b* for a number of values of Q, as indicated in the figure. These values of Q are related to those shown in Figure 1b, with their values of ωL/R when all acoustical and mechanical values have been transformed to the electrical domain.

The final equivalent circuit is shown in Figure 1-8c, with the mechanical and

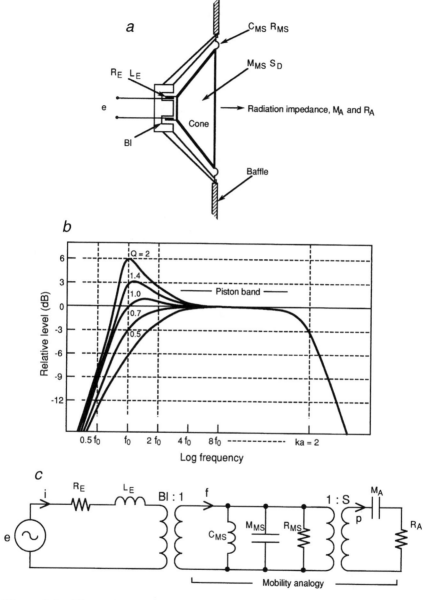

Figure 1-8. The loudspeaker driver: a simple electro-mechano-acoustical transducer. Physical view (*a*); typical ranges of response (*b*); equivalent electrical circuit with mobility analogy in mechanical and acoustical portions (*c*).

acoustical elements shown in the mobility analogy. The terms used in the equivalent circuit are known as the *electromechanical parameters*.

1.6.1 Loudspeaker Efficiency and Sensitivity

For typical loudspeakers intended for LF applications, the range between f_0 and $ka = 2$ may be about a frequency decade (10-to-1 range), or slightly more. The frequency range of flat output for a loudspeaker placed in a large baffle is often called the *piston band* of the loudspeaker, and we can easily calculate the efficiency of the device in this range, as measured on one side of a large baffle. This is the so-called half-space efficiency.

Efficiency is the measure of the acoustical power output divided by the electrical power input to the loudspeaker. At any given frequency in the piston band, the overall efficiency is the product of the electrical-to-mechanical and mechanical-to-acoustical conversion efficiencies. As we have seen, the balance between these quantities is not constant with frequency, but their product is. In the frequency range in which the motion of the cone is mass controlled, and in which the directivity of the loudspeaker is uniform, Small (1972) gives the conversion efficiency, η_0, of the loudspeaker as:

$$\eta_0 = [\rho_0(Bl)^2 S_D^2]/[2\pi c R_E M_{MS}^2] \qquad (1.14)$$

where:
    B = magnetic flux density (T)
    l = length of voice coil (m)
    ρ_0 = density of air (1.18 kg/m^3)
    R_E = voice coil resistance (Ω)
    M_{MS} = total moving mass (kg)
    S_D = area of diaphragm (m^2)

As an example, let us consider the JBL Model 2226H driver. The relevant parameters are given as:

Bl = 19.2 T-m
R_E = 5 Ω
M_{MS} = 0.096 kg
S_D = 0.088 m^2

Calculating:

$$\eta_0 = [1.18(368.7)(7.74) \times 10^{-3}]/[2166.6(5)9.2 \times 10^{-3}]$$
$$\eta_0 = 0.0338, \text{ or about } 3.4\%$$

The measured efficiency of the driver is 3.3%, and this is excellent agreement. If we know the efficiency of a driver we can determine its sensitivity. Sensitivity is normally stated as the sound pressure level measured at a distance of 1 m on-axis with a power input of 1 W to the driver, with the driver mounted in a large baffle so that it radiates effectively into half-space.

From physical acoustics we have the following equation for determining the sound pressure level (dB) at a distance r from a sound source in a free-field radiating power W (Beranek, 1954, p. 314):

$$dB\ Lp = 10\ \log W + 10\ \log \rho_0 c + 94 + 10\ \log (1/4\pi r^2) \qquad (1.15)$$

where r is the distance from the source.

With 1 electrical W applied to the JBL 2226H transducer, we know that the output power will be 0.033 W. With r taken as 1 m:

$$dB\ Lp = 10\ \log (0.033) + 10\ \log (1.18 \times 345) + 94 + 10\ \log (0.08)$$
$$= -14.8 + 26.1 + 94 - 11 = 94.3\ dB$$

If the power is radiated into half-space, then the pressure will be 3 dB greater, giving a total of 97.3 dB. The published sensitivity value for the 2226H is 97 dB, so the agreement is excellent.

We can construct a simple table for converting efficiency directly into 1-W, 1 m piston band sensitivity:

Half-space efficiency	1-W, 1-m piston band sensitivity
25	106
20	105
16	104
12.5	103
10	102
8	101
6.3	100
5	99
4	98
3.15	97
2.5	96
2	95
1.6	94
1.25	93
1	92

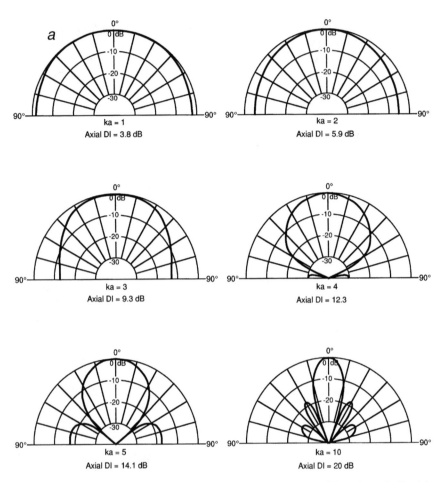

Figure 1-9. Directional (polar) response of a piston mounted in a large baffle (*a*); directional response of an unbaffled piston (*b*); directional response of a piston mounted in the end of a long tube (*c*).

1.7 Directional Characteristics

We have almost finished our model. Our final step is to consider the directionality of the radiating surface and how it might extend the useful frequency range of a loudspeaker. A circular piston mounted in a large wall will exhibit a radiation pattern that is highly dependent on frequency and the angle of observation. for an observer on axis at some distance from the piston, radiation from all portions of the piston will arrive virtually at the same time, and the response will be maximum for all frequencies. For an off-axis position, radiation from some portions of the piston will arrive later than others, and the response will be less

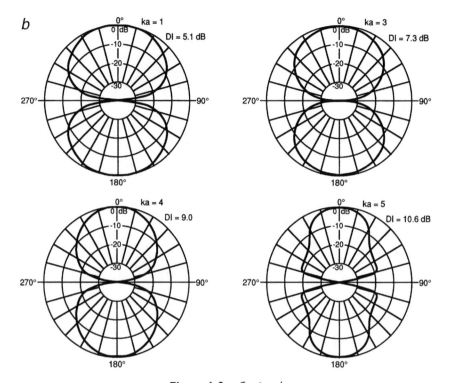

Figure 1-9. *Continued*

than that observed on axis, especially at higher frequencies (shorter wavelengths). This is intuitively obvious, and the effects are shown in Figure 1-9.

Polar plots for a wall-mounted piston are shown in Figure 1-9a. For wavelengths that are long with respect to piston radius, the directionality will be virtually uniform. For shorter wavelengths, it will become progressively more directional on axis. The scale factor here is the quantity *ka*

where:
k = wave number, $2\pi/\lambda$
a = radius of the piston (m)

The directional function, $\Gamma(\alpha)$, is given by:

$$\Gamma(\alpha) = 2J_1(ka \sin \alpha)/ka \sin \alpha \qquad (1.16)$$

where:
J_1 represents a Bessel function of the first order
a = radius of piston (meter)

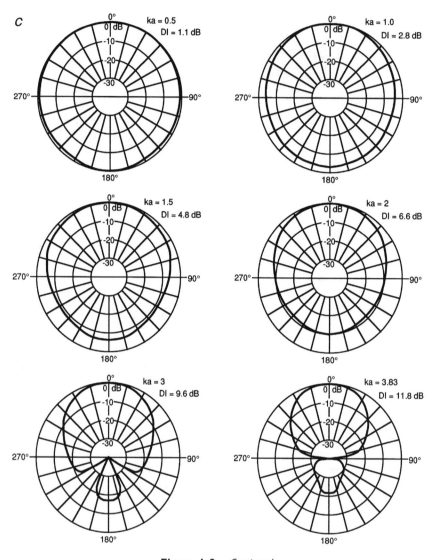

Figure 1-9. *Continued*

λ = wavelength
α = off-axis measurement angle

As a convenient rule, recall that *ka* equals circumference divided by wavelength. For values of *ka* of 1 or less, we can assume that the radiation is uniform over the solid angle in front of the large wall. For values of *ka* of 3 and greater, the piston takes on progressively more pronounced directivity.

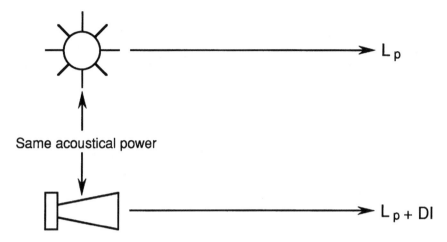

Same acoustical power

Figure 1-10.   A graphical definition of directivity index (DI).

Directionality for an unbaffled piston (a dipole) is shown in Figure 1-9*b* and for a piston mounted at the end of a long tube in Figure 1-9*c*.

A convenient way to quantify the on-axis directional response of a loudspeaker is by *directivity index* (DI). DI is the ratio in dB of sound pressure radiated along the preferential axis of a device, as compared to the sound pressure at the same measuring point if all power from the device were radiated omnidirectionally. A graphical definition of DI is given in Figure 1-10.

1.7.1 The Final Model

As we have seen, the range of flat power response for our driver model extends from the fundamental resonance, f_0, up to $ka = 2$, the frequency at which the power response begins to roll off. In practice, we can use a loudspeaker above that point, taking advantage of the on-axis increase in output due to directional effects. As a rule, we can just about double the upper usable frequency, allowing it to attain a DI of about 10 dB, corresponding to $ka = 3$. This corresponds roughly to a 380-mm (15-in.) loudspeaker maintaining a reasonably flat on-axis response to approximately 1 kHz.

Figure 1-11 summarizes the data given in Figures 1-7, 1-8, and 1-9 in terms of on-axis directivity index as a function of ka.

1.8 Cone Excursion, Power, and Pressure Relationships in Graphical Form

Figure 1-12 shows the relationship of power output from a piston on one side of a large baffle as a function of amplitude, piston radius, and driving frequency. The governing equation is:

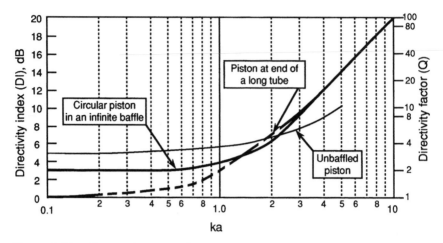

Figure 1-11. On-axis directivity of a piston in a large baffle, at the end of a long tube, and in free space.

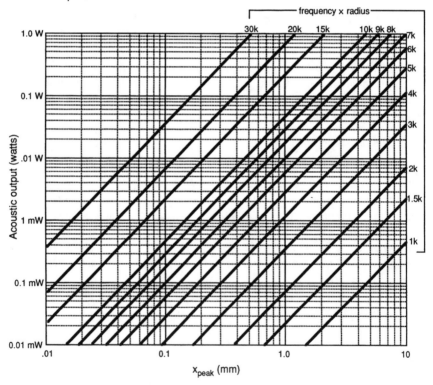

Figure 1-12. Acoustical power output on one side of a piston mounted in a large baffle as a function of amplitude, radius, and frequency.

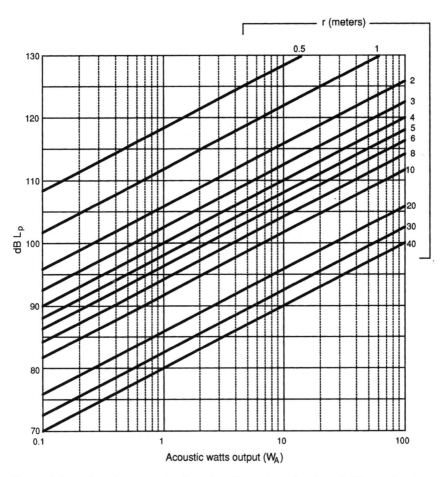

Figure 1-13. Sound pressure level produced by a piston in a large baffle as a function of radiated power and distance.

$$W_A = [x_{peak}(f^2a^2)]^2 / 2.32 \times 10^{17} \qquad (1.17)$$

where W_A is the radiated power, x is the peak amplitude of motion (mm), f is the driving frequency, and a is the piston radius (mm).

Figure 1-13 shows the sound pressure level produced by a piston in a large baffle as a function of radiated power and distance. The governing equation is:

$$L_p = 112 + 10 \log W_A - 20 \log r \text{ (dB)} \qquad (1.18)$$

where W_A is the radiated power and r is the distance (m).

Bibliography

Beranek, L., *Acoustics*, Wiley, New York (1954).

Borwick, J., *Loudspeaker and Headphone Handbook*, Butterworths, London (1988).

Colloms, M., *High Performance Loudspeakers*, 4th ed., Wiley, New York (1991).

Eargle, J., *Electroacoustical Reference Data*, Van Nostrand Reinhold, New York (1994).

Hunt, F., *Electroacoustics: The Analysis of Transduction and its Historical Background*, Acoustical Society of America, New York (1982).

Kinsler, L., et al., *Fundamentals of Acoustics*, 3rd ed., Wiley, New York (1982).

Massa, F., *Acoustical Design Charts*, The Blakiston Company, Philadelphia (1942).

Olson, H., *Acoustical Engineering*, D. Van Nostrand, New York (1957).

Rice, C., and Kellogg, E., "Notes on the Development of a New Type of Hornless Loudspeaker," *Transactions, AIEE*, Vol. 44 (1925), pp. 461–475.

Sakamoto, N., *Loudspeakers and Loudspeaker Systems*, Nikankogyu Shimbun, Tokyo (1967) (in Japanese).

Siemens, E., U. S. Patent 149,797, issued April 14, 1874.

Small, R., "Direct Radiator Loudspeaker System Analysis," *J. Audio Engineering Society*, Vol. 20, No. 5 (1972).

Cone and Dome Drivers

2.1 Introduction

Cone and dome drivers form the backbone of loudspeaker system design, primarily because of their ease and economy of manufacture and their high level of performance. In this chapter we will examine these factors in detail, as well as discussing many variations on the basic design theme. Mechanical performance limits and distortion will also be addressed.

2.2 Mechanical Construction Details of the Cone Driver

Figure 2-1 shows a section view of a 380-mm (15-in.) cone driver with the essential parts labeled. We will use this figure as a guide in discussing the design and construction of the device.

2.2.1 The Frame

The frame, often referred to as the "basket," is the ribbed conical structure that holds the driver together. Professional drivers use frames made of die-cast aluminum, while many lower-cost frames are made of stamped metal. Injection molded plastic is also used. The advantages of the die-cast structure are dimensional accuracy and relative freedom from warping when installed in an enclosure.

While there would seem to be little to comment on regarding frame design, it is worth mentioning that considerable engineering effort has been expended over the years in reducing the metal content in frames while maintaining structural integrity. Thin-wall castings preserve mechanical strength in a manner not unlike stamped automotive body parts. Heat sinking fins and flat surfaces for mounting transformers and/or terminals are among the details that are often designed into frames.

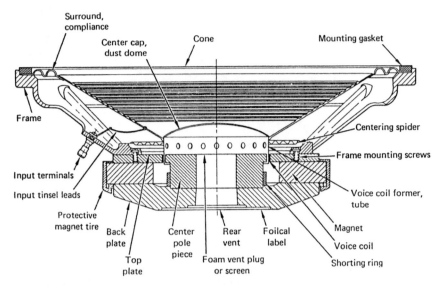

Figure 2-1. Section view of a professional-grade 380-mm- (15-in.) diameter LF driver. (Data courtesy JBL, Inc.)

2.2.2 The Magnetic Motor Structure

Chapter 3 will deal with the engineering specifics of magnetics; here we will consider only the mechanical nature of the magnetic structure and its basic function of providing a uniform magnetic field for the voice coil. In the days before high-energy permanent magnet materials, driver magnets were electrically energized with direct current (dc) flowing through the windings of a field coil. Today permanent magnets are universally used.

Three types of common magnetic motor structures are shown in Figure 2-2. The Alnico V (aluminum-nickel-cobalt) version shown in Figure 2-2*a* was in common use from the period after World War II until the mid-seventies, when cobalt became a very expensive commodity. A typical ferrite, or ceramic, magnet structure is shown in Figure 2-2*b*, and is by far the most common structure today. The structure shown at *c* illustrates the use of high-energy, radially charged small magnets made of rare earth materials such as samarium and neodymium.

Figure 2-3 shows the voice coil-magnetic gap choices available to the transducer engineer. The underhanging voice coil, shown in Figure 2-3*a*, lies completely within the axial range of uniform magnetic flux density. Such a structure is expensive, due to the metal and magnet requirements, but it accommodates a lightweight voice coil (hence a lightweight moving system) and is commonly used in high linearity drivers of moderate to high efficiency.

The form shown in Figure 2-3*b* concentrates all of the flux in the coil at its

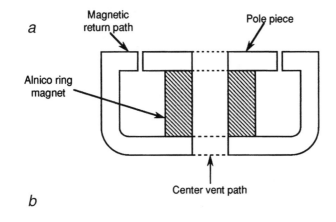

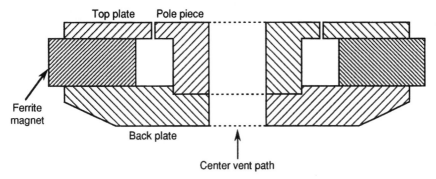

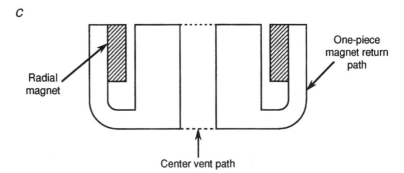

Figure 2-2. Typical magnet-structures. Alnico ring magnet structure (*a*); ferrite magnet structure (*b*); radial magnet structure for high-energy magnet materials.

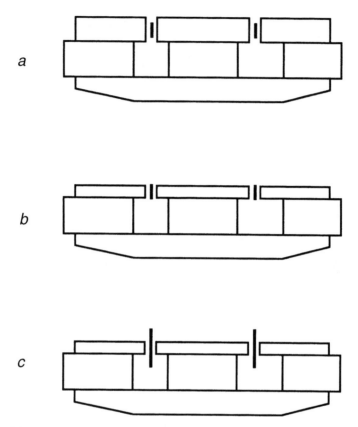

Figure 2-3. Voice coil and top plate topology. Underhanging voice coil (*a*); voice coil and top plate of equal length (*b*); overhanging voice coil (*c*).

rest position. It is evident that even moderate excursions of the voice coil will result in some loss of total flux engaging the voice coil, thus producing distortion. This design is common in very-high-efficiency drivers used for musical instrument amplification, where some degree of distortion may indeed be sonically beneficial.

The overhanging design shown in Figure 2-3c provides for constant flux engaging a portion of the voice coil over a fairly large excursion range. The compromise here is that a large percentage of the voice coil lies entirely outside the flux field at all times, and this constrains the $(Bl)^2/R_E$ ratio to be diminished. This design approach may be applicable for drivers intended for high linearity, but with relatively low efficiency.

The design shown in Figure 2-4 combines elements of the over-and underhanging approaches shown in Figure 2-3. The undercut pole piece is extended above and below the top plate and thus spreads the magnetic flux over a larger axial

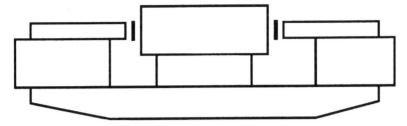

Figure 2-4. Detail of extended pole piece.

range. This design is also beneficial in that it provides an effective heat sinking surface adjacent to the voice coil over a large excursion range.

The motor design shown in Figure 2-5 provides two reversely polarized gaps. A dual voice coil structure with reversely wound coils is used. The push-pull arrangement reduces second harmonic distortion and affords a uniform magnetic field over a fairly large axial range. Another advantage here is that the magnetic field of the motor is effectively contained, producing little stray flux.

Dual voice coils have also been used for combining stereo LF signals in a single driver for systems with a common bass unit. In this application, it is desirable for the coupling between the two windings to be minimal so that amplifier stability does not become a problem.

2.3 The Moving System

The driver moving system consists of five elements: the cone, voice coil assembly, outer suspension (surround), inner suspension (spider), and dust dome, as shown in Figure 2-1. The purpose of the two suspensions is to ensure that the motion of the assembly is basically constrained to the axial dimension, with minimal radial and rocking motions. Each of these elements has a profound effect on the response of the driver, and we will discuss each in detail.

2.3.1 The Cone

The cone is the part of the driver that radiates sound and is the approximation of the ideal piston we discussed in Chapter 1. For operation in the driver's piston band at moderate driving levels, a simple cone made of a light, rigid material does very well. But for large excursions over a wide frequency range there are other considerations.

At high excursions the cone has a tendency to break up, producing erratic response and distortion. At the same time, high-frequency (HF) response of any magnitude of cone motion brings into play complex vibrational modes. The cone profile and material are critical factors in controlling these effects.

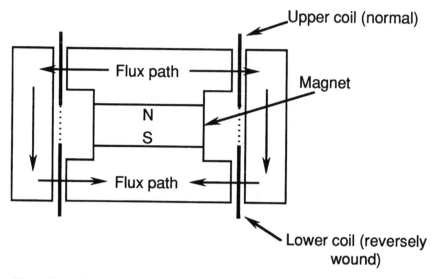

Figure 2-5. Dual gap, with reverse flux paths and reversely wound dual voice coils.

Section views of several cone profiles are shown in Figure 2-6. A shallow cone with annular ribs is shown in Figure 2-6*a*. The purpose of the ribs is to add stiffness to the cone with minimal addition of mass. This promotes good piston action with extended high frequency response. A deep cone is shown in Figure 2-6*b*. In general, a deep cone has greater rigidity than a shallow one, all else being equal, and we find them used in robust designs intended for high-level application. The on-axis HF response of the deep cone may be compromised to some extent due to the fact that rear portion of the cone (the apex) is displaced from the edge of the cone by an amount that will cause a cancellation in response on-axis when the front-to-back distance is equal to a half-wavelength. A shallower cone raises the frequency at which this cancellation will take place.

The so-called curvilinear cone profile shown in Figure 2-6*c* is often used in lightweight moving systems where efficiency and extended output at high frequencies are the chief considerations. The shape of the cone actually promotes decoupling from the voice coil at high frequencies, increasing the driver's output while maintaining a fairly smooth response on axis. The original JBL 380-mm-diameter D-130 from the late 1940s was the archetype of this design, and the approach is still favored in drivers used with electronically amplified instruments.

Figure 2-7 shows typical "bell modes" that a cone exhibits at high frequencies; these modes can often be identified as such in the on-axis response of drivers driven at any level. The figures represent a "snapshot" of a cone, as seen from the front, with the plus and minus signs indicating instantaneous outward and inward motions of the cone.

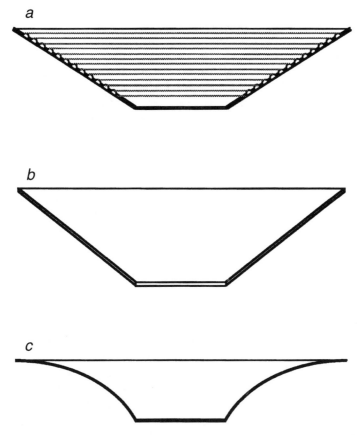

Figure 2-6. Typical cone profiles in section view. Ribbed cone (*a*); straight-sided cone (*b*); "curvilinear" cone (*c*).

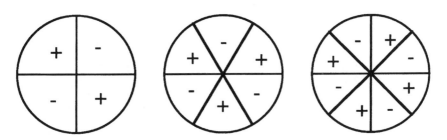

Figure 2-7. Representation of cone bell modes of vibration.

The on-axis data shown in Figure 2-8 shows the effects of cone breakup for three drivers, each designed for a specific application. The JBL 2220 has a light curvilinear cone, and its midband sensitivity of 101 dB, 1 W at 1 m, indicates a piston band efficiency of about 8%. Note that the response extends well beyond 2 kHz with relative freedom of sharp peaks and dips. The JBL 2225 driver is designed for general sound reinforcement applications and has a piston band efficiency of about 3.5%. The pronounced peak in cone breakup at about 4 kHz is well outside the normal bandpass of the driver. The JBL 2235 is designed for use in studio monitor loudspeakers and has a piston band efficiency of 1.3%. Although there are a number of ripples in the pattern of breakup modes, they are small and diminish quickly above 2 kHz.

The response of a typical 125 mm (5 in.) midrange driver is shown in Figure 2-9. Note the similar pattern of breakup, in this case transposed upward in frequency by the inverse of the relative dimensions of the drivers.

2.3.2 Cone Materials

Cone materials include paper (cut and seamed), paper pulp (molded or felted to fit), plastic, metal, and all manner of composites. The most generally used material is felted paper pulp made by drawing a fibrous slurry through a fine screen in the shape of the desired cone. The material is then cured, dried, and trimmed to fit. Molded plastic materials are also in wide use and have the virtues of being more consistent, batch to batch, than the felted variety.

The desirable mechanical characteristics of a cone are rigidity and high damping (internal loss). These characteristics are normally opposed to one another in that anything that improves one will generally work to the detriment of the other.

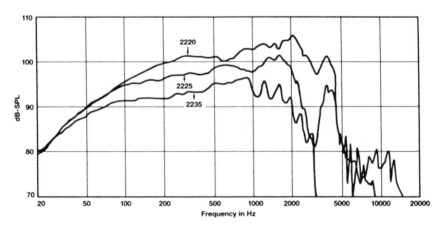

Figure 2-8. On-axis response of JBL Models 2220, 2225, and 2235 LF drivers. (Data courtesy JBL, Inc.)

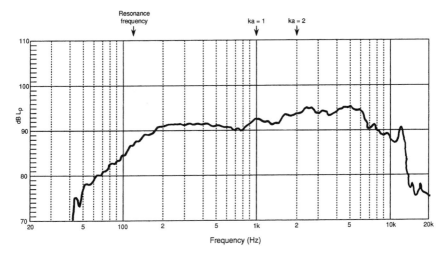

Figure 2-9. On-axis response of JBL Model LE-5 midrange driver. (Data courtesy JBL, Inc.)

The necessity of rigidity is clear; the high excursions that drivers exhibit at low frequencies demand that the cone have sufficient strength not to buckle or deform under the forces imparted to it.

The need for internal damping in the cone is illustrated in Figure 2-10. At sufficiently high frequencies, the cone does not act as a single unit; rather, it acts as a mechanical transmission line with distributed mass, resistance, and compliance, providing a medium for mechanical waves to travel outward as shown. If these waves (shown highly exaggerated) reach the edge of the cone with little attenuation, they will reflect back to the cone apex, creating a pattern of peaks and dips in response.

The cure for this problem lies in a combination of damping in the cone as well as attention to damping in the surround so that it will act, insofar as possible, as a matched load. Figure 2-11 illustrates a typical problem that can occur when a resonance in the surround is driven by an unattenuated signal from the cone. Here, the center section of the surround is vibrating in opposite polarity to the cone, creating what is commonly called a "surround dip" in response.

Figure 2-10. Representation of outwardly traveling waves in a cone.

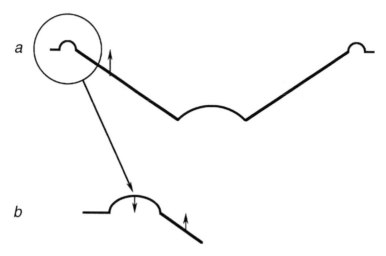

Figure 2-11. Representation of a surround resonance mode in antiphase with cone.

As an example of how well the compromise can be made, we show the on-axis response of the JBL LE-14 driver in Figure 2-12. This device has a felted cone treated on both sides with a controlled amount of damping material known as Aquaplas™. The surround is a half-roll of polyurethane foam that provides an excellent termination for the traveling wave in the cone. Note that the on-axis rolloff in the 1–2 kHz range is very smooth.

Plastic cones are attractive in that the desired attributes of stiffness and damping can be uniformly maintained in production. Additionally, the material is not affected by environmental humidity, as is paper.

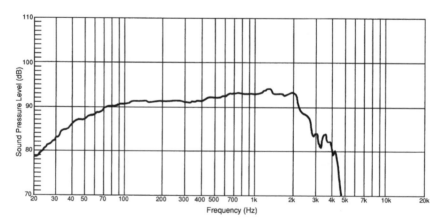

Figure 2-12. On-axis response of JBL model LE-14 LF driver. (Data courtesy JBL, Inc.)

2.3.3 The Outer Suspension (Surround)

Figure 2-13 shows a variety of surround details. The half-roll form shown at *a* is usually made of polyurethane foam. The design offers high compliance and long travel, but the price paid for this is diminished radial stability. The form shown at *b* is a double half-roll, normally made of treated cloth. The design is very rugged and can be treated so that it is fairly stiff and well damped, making it a good choice for drivers intended for music and sound reinforcement.

A common form of mechanical distortion in surrounds is known as *hoop stress*. Imagine that the half-roll surround shown in Figure 2-13a was made of a nonwarping material, such as treated cloth. Then we could view the surround as composed of a number of concentric circular elements, or hoops. Now, as we attempt to move the cone in an axial direction, some of the hoop elements are constrained to move inward (or outward) and thus take on a smaller (or larger) diameter. This of course cannot happen, and instead the surround will tend to wrinkle.

Materials that can warp, such as synthetic rubber or various foams, will actually deform to eliminate hoop stresses, resulting in less radial rigidity in the surround.

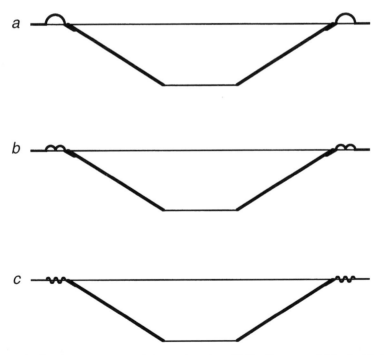

Figure 2-13. Various surrounds in section view. Half-roll surround (*a*); double half-roll surround (*b*); accordion-type surround (*c*).

The multiple roll form shown in *c* offers high excursion capability, but with some tendency for resonances. Olson et al. (1954) present the data shown in Figure 2-14. Here, the nature of the suspension motions are clearly indicated, and the addition of a rubber damping ring to the suspension smooths the response dip, as shown.

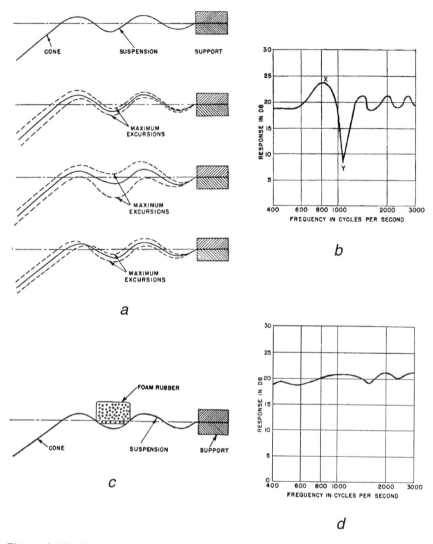

Figure 2-14. Surround resonance modes. Surround and cone motion (*a*); dip in response caused by effect shown in (*a*); (*b*); surround treatment to reduce motion (*c*); improvement in response (*d*). (Data courtesy *Journal of the Audio Engineering Society.*)

2.3.4 The Inner Suspension

Often called the "spider," the inner suspension is connected to the voice coil former, and its outer edge is bonded to the frame. The original designs, many years ago, had legs that were bonded to the frame, hence the familiar name.

Figure 2-15 shows profile views of typical spiders. The form shown in Figure 2-15*a* is a flat spider, and the form shown in Figure 2-15*b* is a cup spider. The material normally used is a fairly open-weave cloth impregnated with phenolic resin and formed under heat. The multiple rolls afford excellent opportunities for spurious resonances at high drive levels, just as we have observed in the outer suspension.

In some driver designs, the spider must be porous in order to provide venting of the volume of air captured behind it. It is clear from the section view shown in Figure 2-1 that venting in that design is through a center hole at the back of the magnet structure. However, many magnet structures are not open to the rear, and venting must take place through the spider or through a porous section in the dust dome.

2.3.5 The Voice Coil

Figure 2-16 shows a drawing of a voice coil that has been directly wound onto the former. The design shown here is made of flat (ribbon) aluminum wire milled specifically for this purpose. The return of the bottom lead on the inside of the former can be seen in the drawing. Both leads exit at the top of the coil and, after the coil assembly has been glued to the cone, they are dressed on the front of the cone to a set of eyelets near the apex of the cone. From here, they are connected by loops of tinsel lead on the back side of the cone to a set of terminals on the frame.

In drivers intended for professional applications, ribbon wire has several distinct advantages over round wire:

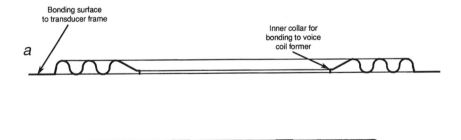

Figure 2-15. Section view of spider. Flat type (*a*); cup type (*b*).

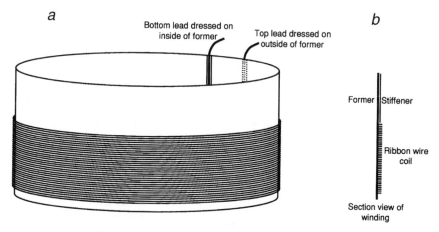

Figure 2-16. Drawing of an assembled voice coil.

1. Greater packing density. Theoretically, the dead space between adjacent windings is minimal, providing about a 27% improvement over round wire, providing minimum R_E for the volume occupied.

2. Slightly increased rigidity. The tight bonding between adjacent layers of the coil provides an integral structure.

3. Impedance adjustment. This is by far the most significant advantage of ribbon wire, as shown in Figure 2-17. It is possible, by adjusting the ribbon thickness appropriately, to operate at a new design impedance with no change in the $(Bl)^2/R_E$ ratio.

In Figure 2-17a, assume that we have a voice coil at a design impedance of, say, 8 Ω. We wish now to design a new coil that presents a 16Ω impedance. To do this, we mill the wire so that it has 0.7 the thickness, as shown in Figure 2.17b. It is clear that the total length of wire in the voice coil will now be 1.4 times as in Figure 2.17a. The new R_E will be twice what it was at a (because of the added length and reduced cross section), and the new value of Bl will be 1.4 times greater. However, $(Bl)^2$ will increase by a factor of 2, so we have kept the same ratio as before.

Voice coils are invariably made of aluminum or copper. The resistivity of aluminum is about 1.6 times that of copper, but its advantage lies in its lower mass, which is less than one-third that of copper.

Typical impedance curves for 8- and 16-Ω versions of the same driver are shown in Figure 2-18. In both cases the driver was mounted in a 280-L sealed enclosure. The rise in impedance at the fundamental system resonance, as modified by the enclosure volume, is apparent and is a reflection of the mechanical

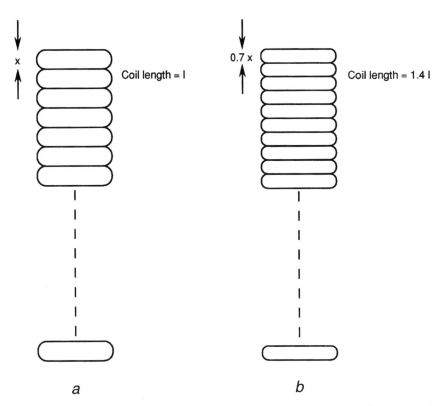

Figure 2-17. Voice coil impedances. Section view showing flat wire profile for normal impedance coil (*a*); section view showing flat wire profile for two-times normal impedance (*b*).

resonance through the *Bl* coupling coefficient into the electrical domain. The rise in impedance at higher frequencies is due to the inductance of the voice coil, L_E. It is the dominant element in the driver's impedance above about 1 kHz.

The minimum value of impedance in the region just above the impedance peak is where all reactive terms have canceled each other, and we are left only with resistive terms: voice coil resistance, mechanical loss terms reflected through to the electrical domain, and of course the radiation load itself.

The effect of voice coil inductance can be partially reduced by plating a highly conductive copper or silver ring on the pole piece in the vicinity of the voice coil. In this application, the plated ring acts as a shorted turn secondary winding of a transformer and swamps out the effect of the inductance. This is commonly done in any large driver that is designed for both mid-frequency (MF) and HF response.

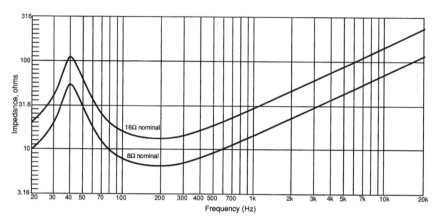

Figure 2-18. Typical impedance curves for 8- and 16-Ω versions of the same LF driver. Impedance shown on logarithmic scale.

2.3.6 The Dust Dome

Almost an afterthought, the dust dome is a spherical section, normally of molded paper or stamped aluminum, that is glued to the top of the voice coil former as the driver reaches the end of the assembly line. Its primary purpose is to keep debris from entering the magnetic gap, but it also functions to stiffen the moving system somewhat.

Over the years, numerous "full-range" drivers have been designed with aluminum domes that take advantage of the decoupling of the relatively massive cone from the voice coil at high frequencies. On-axis response can be maintained to fairly high frequencies, purely through the direct mechanical coupling of the voice coil to the aluminum dome.

2.4 Variations on the Cone Transducer

In its nearly three quarters of a century of existence, the cone transducer has seen a vast number of modifications and changes, and in this section we will describe some the more interesting of these modifications.

2.4.1 The Decoupled Cone

As we have stated, in all cone drivers there is a certain amount of decoupling of the voice coil and cone at high frequencies. One of the earliest efforts here was the Duo-Cone design of the Altec-Lansing Company, as shown in Figure 2-19. The cone was divided about midway, providing it with an added suspension element and damping for that element. The assumption was that, at low frequen-

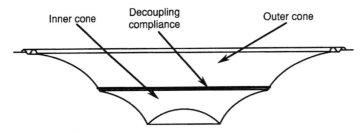

Figure 2-19. Section view of the Altec-Lansing Duo-Cone profile.

cies, the entire cone moved as a unit, while at higher frequencies only the inner, smaller, cone was moving. In many ways, such a design as this tended to elevate loudspeaker driver manufacturing to the level of making a violin! Fine in concept, but very difficult to control on the production line.

2.4.2 The "Whizzer" Cone

To take full advantage of the decoupled voice coil, a free-edged whizzer cone can be attached to the voice coil former, providing a highly resonant acoustical termination for direct radiation of very high frequencies, as shown in Figure 2-20. Whizzers abound today in low-priced loudspeaker systems and actually perform better than one would suspect. Again, there is considerable response variation from unit to unit.

2.4.3 The JBL LE-8

The venerable LE-8 dates from the early 1950s. This remarkable 200-mm- (8-in.) diameter driver has a light cone, treated with Aquaplas for damping. The top plate is thick and accommodates an underhanging voice coil for good linearity. The 50-mm (2-in.) aluminum dust dome is carefully damped on the underside

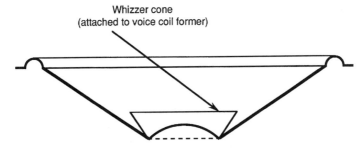

Figure 2-20. Section view showing "whizzer" high frequency (HF) cone attached to LF cone.

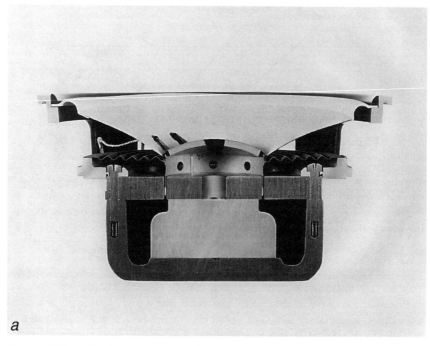

a

Figure 2-21. The JBL model LE-8 full-range cone driver. Photograph (*a*); typical on-axis response (solid curve); dashed curve indicates typical 45° off-axis response (*b*). (Data courtesy JBL Inc.)

to control its response. Figure 2-21*a* shows a photo of the LE-8, and typical response is shown in Figure 2.21*b*. The LE-8 was once very popular as a close-in monitor loudspeaker, where its axial response predominates over its power response, which falls off rapidly at high frequencies. For controlled response beyond 10 kHz, the LE-8 required more on-line manufacturing quality control than did the Duo-Cone design.

2.4.4 Dual Magnetic Gap Designs

The idea of combining two or more transducers on one chassis has always been attractive, and there are numerous successful designs attesting to this. None has been as elegant as the British Tannoy Dual-Concentric, which uses a single magnet with two parallel gaps, one for the low frequency section and the other for the horn-loaded high-frequency (HF) section. A section view of the magnetic structure is shown in Figure 2-22.

Another variation here is the induction HF system, where a full-range signal

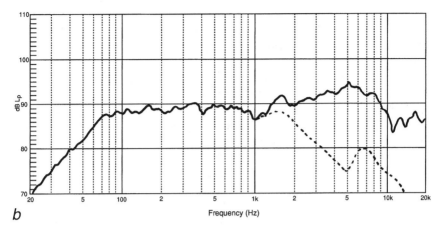

b

Figure 2-21. *Continued*

is fed to the LF coil, and by induction, a second coil located close to it drives a small radiating surface at high frequencies.

2.5 Dome Drivers

It is very difficult to make a small-cone HF driver with double suspensions because of size limitations, and the simple dome driver has become the standard

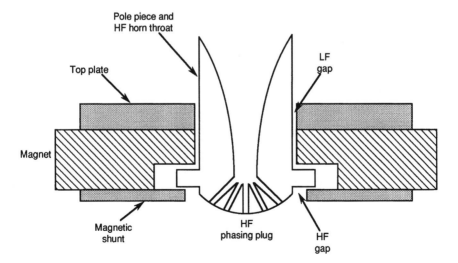

Figure 2-22. Section view of the Tannoy Dual-Concentric loudspeaker magnetic structure.

here. The major structural difference between the cone and dome is that the dome has only a single suspension. The dome resembles a very small cone loudspeaker with its cone removed, leaving only the inner suspension, voice coil, and the dust dome operative.

Figure 2-23 shows a section view of a typical HF (25-mm) dome. There is normally a metal mesh screen on the front to protect the dome. Typical materials used for the dome construction include metals such as aluminum, titanium, and beryllium. Soft domes are made of thin molded cloth material that has been sized and "doped" with a viscous damping compound.

Although the metal dome is a segment of a sphere, the radiation from the dome is not that of a pulsating sphere. Rather, it moves as a unit and will exhibit a pattern of HF dips in on-axis response as portions of it propagate in and out of polarity with each other. This arrangement is shown in Figure 2-24. For example, a dome with height of *h* meters will show a dip in response at the frequency whose wavelength is equal to *h/2*. For a dome HF device with a height of 10 mm:

$$f = 345/[2(10) \times 10^{-3}] = 17.25 \text{ kHz} \tag{2.1}$$

While such a response dip is normally located at the upper end of the device's bandpass, it is worth noting that the phase loss is a gradual one and may be noticeable in the 5- to 6-kHz range.

The typical soft dome acts as a unit at lower frequencies; however, at progressively higher frequencies, the mass of the dome gradually decouples so that at the highest frequencies the radiation is primarily from the voice coil itself. It in effect becomes a ring radiator at the highest frequencies.

Because of the decoupling, the radiating mass is small at high frequencies, and the output power can be maintained fairly flat. The high internal damping in the moving system tends to flatten the overall response, freeing it of prominent peaks and dips. The basic shortcoming of the typical soft dome is its fairly modest power rating.

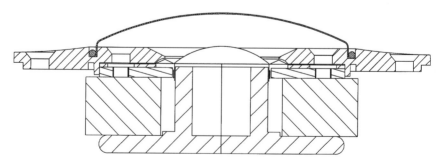

Figure 2-23. Section view of a typical dome HF driver. (Data courtesy JBL, Inc.)

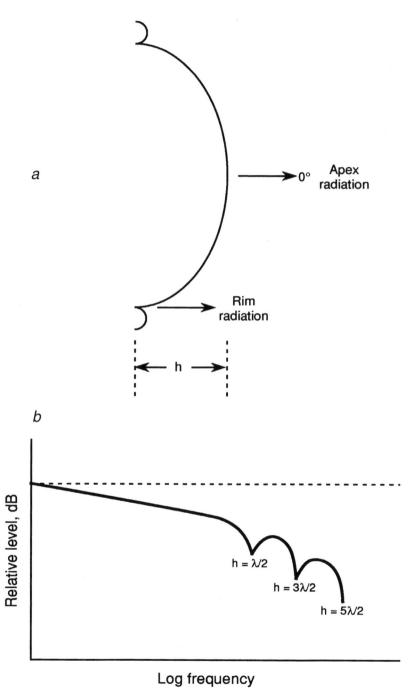

Figure 2-24. On-axis HF losses in a done transducer.

Well-designed metal and soft dome high-frequency devices normally exhibit flat on-axis response out to 20 kHz, or slightly beyond that frequency.

2.6 Distortion in Cone and Dome Drivers

In general, the very low mechanical driving point impedance of the voice coil and motor system will swamp out many small nonlinearities in the suspension elements of a driver. In a well designed driver, it is only at high excursions that suspension nonlinearities become significant.

Another source of distortion in the driver may come as a result of nonuniform magnetic flux surrounding the voice coil as it makes its excursions. Yet another form of distortion is due to the particular interaction of voice coil current with the static flux field provided by the magnet itself. We will examine some of these effects.

2.6.1 Mechanical Nonlinearities

The nominal power rating of a driver is based on long-term effects of heating as well as the effects of excessive displacement at lower driving frequencies. Many 380-mm (15-in.) drivers intended for heavy-duty professional applications may have a nominal power rating of 600 W, implying that the device can withstand continuous power input of that amount with a given input spectrum.

The JBL Model 2226H driver whose response is shown in Figure 2-25 is rated at 600 W with a shaped pink noise signal over the frequency decade from 50 to 500 Hz. The noise signal is limited to a 6-dB crest factor. Normal operation of

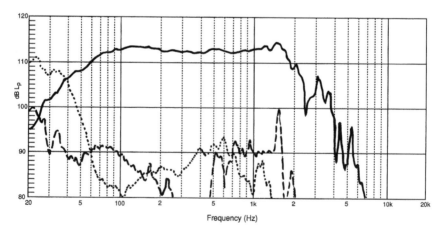

Figure 2-25. On-axis response of JBL model 2226 LF driver, with second (dotted curve) and third (dashed curve) harmonic distortion. Distortion raised 20 dB. (Data courtesy JBL, Inc.)

the driver is in the range perhaps 10 dB lower than full rating (one-tenth rated power), so it is customary for the manufacturer to publish the 60-W distortion data for such a driver.

Generally, second and third harmonic components are dominant in the driver's distortion signature. Higher harmonic components are usually indicative of some gross defect, such as a rubbing voice coil or rattle. In data presentation it is customary to raise the distortion components 20 or 30 dB, as the case may be, so that they can be read on the same graph as the fundamental.

In the case of the JBL 2226H we can see that the reference fundamental output in the piston band is about 113 dB Lp, as measured at 1 m. Over the range from 60 Hz to about 600 Hz, the levels of second and third harmonics are about 40 dB below the fundamental (recall the 20-dB offset in the graph), and this corresponds to distortion values for each harmonic of about 1%. At low frequencies the third harmonic dominates, due to the nonlinearity of both inner and outer suspensions at their excursion limits, as well as the loss of Bl product at excursion extremes. Above 1 kHz, a second harmonic peak of −33 dB appears and is associated with breakup modes of the cone.

The distortion generated by a 25-mm (1-in.) titanium dome device is shown in Figure 2-26. Again, the curves are run at one-tenth the nominal power rating of the device. Most of the distortion here comes as a result of breakup modes in the diaphragm.

Another form of distortion found in some LF drivers is called *dynamic offset*. This is a tendency for the cone to execute a displacement from its normal rest position during high-level LF signals. This static displacement can be either in or out of the driver. While the actual cause of the effect is rather complex, it can be roughly described as follows. During high excursions there may be a lowering of *Bl* product as the voice coil leaves the gap, during which times the motional impedance of the driver drops. When this happens the driver will draw greater current from the amplifier. Since the driver's mechanical elements are never quite symmetrical, there will be a tendency for the cone to pull to one side or the other during such heavy drive signals. The general cure for dynamic offset is to use inner suspensions whose stress-strain curve is matched to the loss of Bl product, thereby offering more resistance to cone displacement as the voice coil moves out of the gap.

Another type of distortion sometimes found in drivers is *subharmonic generation*. This normally happens in very lightweight cones or domes when they are driven at very high levels. Portions of the cone or dome will execute alternate in-and-out motions as they are excited by driving frequencies an octave higher. While these effects are quite audible on sine wave test signals, they are not normally audible on program material. The cure for subharmonic generation is the use of stiffer cone and dome materials.

Rocking modes of cone or dome vibration may become apparent at high drive levels at specific frequencies. During such times the acoustical output usually

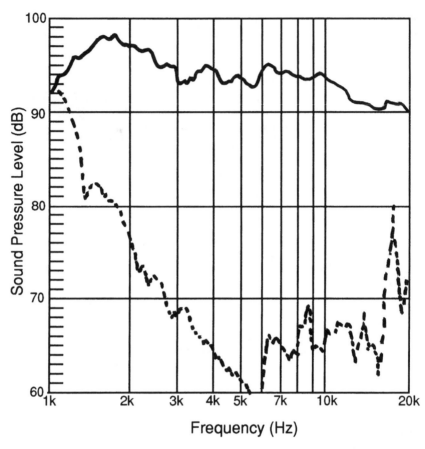

Figure 2-26. On-axis response and 2nd harmonic distortion of a 25-mm (1-in.) titanium dome HF unit at one-tenth rated power. (Data courtesy JBL, Inc.)

drops, and the voice coil may have a tendency to rub against the top plate or pole piece. Extended operation under these conditions will lead to failure of the voice coil. In some rare cases, the condition is influenced by standing wave patterns in the loudspeaker enclosure, where the acoustical loading may vary from one portion of the cone to the next.

2.6.2 Distortion Due to Magnetic Nonlinearities

Example of the distortion generated by magnetic flux modulation are shown in Figure 2-27. The same moving system was used in making the measurements; only the magnetic structures were changed. The data shown in Figure 2-27*a* is for an earlier design Alnico V magnet structure. The distortion is about 1% over the range from 100 Hz to 1 kHz. The data shown in Figure 2-27*b* is for a

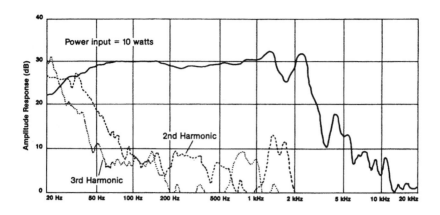

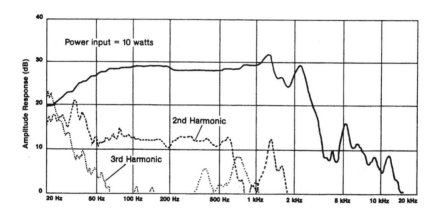

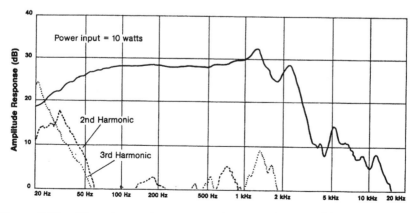

Figure 2-27. Effect of magnetic structure on harmonic distortion. Same moving system mounted in Alnico V structure (*a*); ferrite structure (*b*); and Symmetrical Field Geometry™ structure (*c*). (Data courtesy JBL, Inc.)

45

conventional ferrite magnet structure, showing a moderate amount of second harmonic distortion in the midrange resulting from flux modulation.

The data shown in Figure 2-27c illustrates a ferrite structure that has been fitted with a flux-stabilizing ring as well as further changes in the pole piece geometry. Note that the distortion has been reduced to well below 1% in the range from 100 Hz to 1 kHz. The specific means by which the flux stabilizing ring works will be discussed in Chapter 3 which deals with magnetic systems.

Bibliography:

Beranek, L., *Acoustics*, McGraw-Hill, New York (1954).

Borwick, J., ed., *Loudspeaker and Headphone Handbook*, Butterworths, London (1988).

Colloms, M., *High Performance Loudspeakers*, 4th ed., Wiley, New York (1991).

Henricksen, C., "Loudspeakers, Enclosures, and Headphones," in Ballou, G., ed., from *Handbook for Sound Engineers*, Sams, Indianapolis, IN (1987).

Jordan, E., *Loudspeakers*, Focal Press, London (1963).

McLachlan, N., *Loudspeakers*, Dover, New York (1960).

Olson, H., *Acoustical Engineering*, Van Nostrand, New York (1957).

Olson, H., et al., "Recent Developments in Direct Radiator High Fidelity Loudspeakers," *J. Audio Engineering Society*, Vol. 2, No. 4 (1954).

Sakamoto, N., *Loudspeakers and Loudspeaker Systems*, Nikankogyo Shimbun, Tokyo (1967) (in Japanese).

Various authors, "*Loudspeakers*, Vols 1 and 2, anthologies of papers from the pages of the *Journal Audio Engineering Society*, New York (1978, 1984).

Various authors, "Choosing JBL Low-Frequency Transducers," Technical Note, Vol. 1, No. 3, JBL, Inc., Northridge, CA (1983).

Principles of Magnetics

CHAPTER

3

3.1 Introduction

In this chapter we will cover basic magnetic theory, including an overview of the steps involved in designing a typical magnet structure for a cone driver. We will also discuss aspects of materials and examine some of the fundamental nonlinearities in the magnetic and moving systems.

3.2 Fundamentals

We can create a magnetizing field by passing direct current through a coil of wire. In the centimeter-gram-second (cgs) system of units the magnetizing force field, H, is given in oersteds (Oe), and the resulting magnetic induction, B, is given in gauss (G). The H field is proportional to the current and the number of turns in the coil. If we measure the H and B fields as they exist in air (no magnetic sample present in the coil), then the numerical values of H and B will, by definition, be equal. This simple fact is why so much magnetic system design takes place in the cgs system of units rather than in SI [or meter-kilogram-second (mks)] units.

If we now place a ferromagnetic sample in the coil the situation will change dramatically, as shown in Figure 3-1. If the sample is unmagnetized to begin with, we start at position 1 in the diagram. As we increase the magnetizing field, the induction in the sample will follow the curved path up to position 2, where the induction or B field will level off as it reaches saturation.

If we then reduce the H field back to zero we will observe that the B field is only reduced to some value at position 3, known as B_r, or the *remanance* induction. Our sample is now permanently magnetized. Now, if we reverse the flow of current in the coil and increase it, we will eventually reduce the magnetic induction

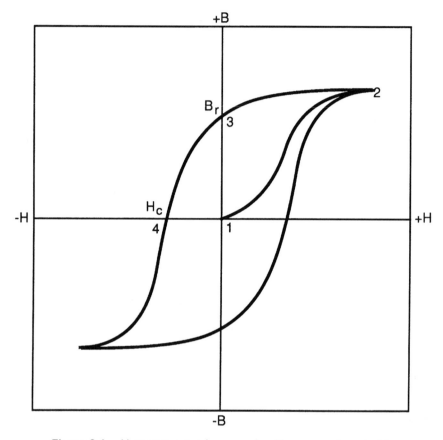

Figure 3-1. Hysteresis curve for a sample of ferromagnetic material.

in the sample to zero at position 4 in the graph. The value of the H field that will cause this is known as H_c, the *coercivity* of the magnetic sample.

3.2.1 Hysteresis

We can continue the process, remagnetizing the sample in the opposite direction, and then demagnetizing it again. The curve that this process outlines is known as a *hysteresis* curve (from the Greek meaning *to lag*).

The second quadrant of the graph (upper left portion) is known as the demagnetizing quadrant, and it is the behavior of magnetic materials in this portion of the graph that will be of greatest interest to us. In Figure 3-2 we will look at several magnetic samples only in the second quadrant, carefully observing their values of B_r and H_c. Curve 1 is for Alnico V material, a common magnetic material widely used from the 1940s to about the mid-1970s, when chronic cobalt

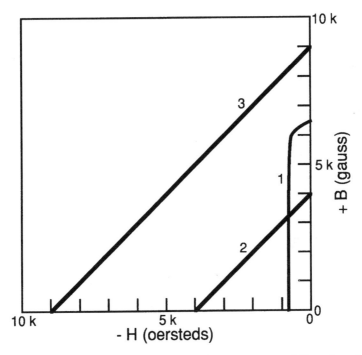

Figure 3-2. Demagnetizing curves for Alnico V (1), ferrite material (2), and neodymium material (3).

shortages forced the loudspeaker industry to find other solutions. The high value of B_r indicates that the material can create high values of flux in a driver, but the low value of H_c indicates that it is relatively easy to demagnetize.

Curve 2 is typical of many ferrite (or ceramic) materials. Such magnet materials are made of various ferrite (Fe_2O_3) combinations with other oxides and are generally far less costly than Alnico V.

As can be seen in the figure, the low value of remanance, B_r, indicates that a large surface area will be required to generate a high flux level in the gap. The high value of H_c indicates that the ferrite magnet is relatively difficult to demagnetize.

Curve 3 is typical of magnets such as the neodymium-iron-boron type that have very high energy for their size and weight. They combine the best properties of both Alnico V and ferrite, exhibiting both high values of B_r ana H_c.

3.2.2 Constructing the Load Line

In designing a magnet structure the transducer engineer targets a specific flux density and the geometry of the magnetic gap over which that value is to exist.

The next step is normally to specify the minimum amount of magnet and iron material that will satisfy these requirements. Figure 3-3*a* shows the second quadrant for a typical sample of Alnico V, and Figure 3-3*b* shows the associated external energy, B_dH_d, associated with it.

When an air gap is placed in the magnetic circuit, magnetic flux passes through the gap, and we are in effect partially demagnetizing the magnet. The load line helps the engineer to get the most efficiency from the magnet by maximizing the energy product, B_dH_d. The engineer draws a horizontal line left from the B_r point and a vertical line upward from the H_c point. The load line is then drawn from the origin (lower right corner of the graph) to the intersection of these two lines, as shown. The intersection of the load line and the demagnetization curve is the target operating point of the system. In Figure 3-3b, we have plotted the product of B and H for all the values in the second quadrant, and a horizontal line drawn from the operating point will intersect the B_dH_d curve at its maximum value. The slope of the load line, B_d/H_d, is known as the *permeance* of the magnet.

The magnetic requirements will be minimized when the BH product is maximized, but for various reasons the engineer may not choose to operate the system at this point. For example, a steeper load line slope will result in an operating point closer on the demagnetization curve to the value of B_r, and operating the system at this new point will make it more immune to demagnetization effects.

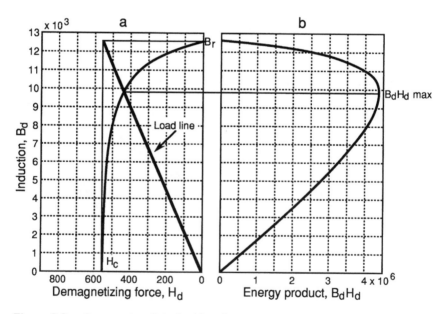

Figure 3-3. Construction of the load line. Demagnetization curve for Alnico V showing load line (*a*); energy product curve for Alnico V (*b*).

This will require a larger magnet, but the choice may be a good one, depending on the intended use of the driver.

3.3 Details of the Magnetic Circuit

The magnetic circuit, commonly called the motor structure, is composed of iron and magnetic material and was discussed in detail in Chapter 2. We may view the relationship between flux and magnetomotive force as an equivalent to Ohm's law for a simple series electrical circuit, where flux (ϕ) is analogous to current, magnetomotive force (M) is analogous to voltage, and magnetic reluctance (R) is analogous to resistance. Thus:

$$\phi = M/R \qquad (3.1)$$

In order to get requisite flux densities in the gap of a transducer, we need an iron material with low enough reluctance and proper gap geometry to concentrate the flux where it is needed. The iron commonly used in the structure is what is called mild steel; it is low in carbon and easily worked. Figure 3-4 shows the *H* field requirements for various amounts of induction for some metals used in magnetic structures. A mild steel material such as EN1A can produce gap flux

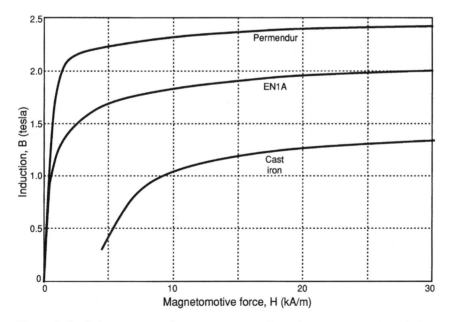

Figure 3-4. Induction curves for two iron materials used in magnet structure design.

up to 1.7 T, which should take care of most transducer requirements. However, for gap requirements in excess of 2 T, as in some compression drivers, a material called Permendur may be needed.

3.4 Linearity Issues

In Chapter 2 we described distortion effects that were caused by nonlinear flux distribution in the gap and by loss of flux during high excursions, both of which can usually be improved through attention to geometry in design. There are other effects involving the voice coil and magnet structure that are not so easily solved, as discussed by Buck (1994):

3.4.1 Flux Modulation

This occurs when the magnet is alternately over-magnetized and reversely magnetized by the cyclic signal current in the voice coil itself. This effectively amounts to a shifting of the instantaneous operating point up and down the demagnetization curve. It is minimal when the demagnetization curve is itself fairly flat, as in the upper portion of a typical Alnico V demagnetization curve.

However, in the case of a ferrite magnet, with its high sloped demagnetization curve, no such solution is available. When JBL, like all other manufacturers, was forced to develop a new ferrite motor structure when cobalt became too expensive, they added a large conductive ring, made of aluminum, at the base of the pole piece. Through transformer action, the voice coil signals that are induced into the magnet structure will set up a countercurrent in the aluminum ring. The ring, a single turn with a cross-sectional area of about 1 cm^2, has an extremely low resistance; consequently, there is substantial current flow through the ring at high signal levels. As in all induction phenomena, the action of the induced current is to counter the effect that produced it in the first place, thus the tendency for flux modulation is greatly reduced at a price about 1 dB of overall magnetic efficiency. A section view of the motor structure with flux stabilizing ring is shown in Figure 3-5.

3.4.2 Interaction Between Voice Coil and Pole Piece

The effect here is that the solenoid action of the energized voice coil magnetically attracts the iron in the pole piece, regardless of the polarity of the signal. This amounts to signal rectification and produces low-level second-harmonic components whose value depends on the *Bl* product of the driver.

3.4.3 Modulation of Voice Coil Inductance

The in-and-out motion of the voice coil over the pole piece varies the amount of iron that is instantaneously inclosed by the voice coil. Since the voice coil

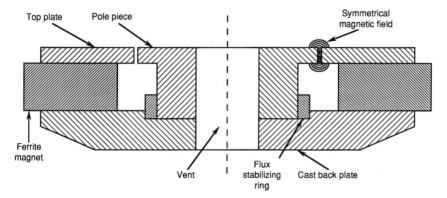

Figure 3-5. Section view of a magnetic structure using an aluminum flux stabilizing ring.

acts as an iron core inductor, the instantaneous inductance of the voice coil changes. The cure for this problem is to plate a silver or copper ring on the pole tip, or affix a copper sleeve to it. This acts as a shorted turn transformer secondary and will reduce the inductance so that its modulation is minimized.

3.4.4 Eddy Currents in the Top Plate and Pole Piece

Current in the voice coil will cause small, local loops of current to flow in the iron adjacent to the voice coil, primarily causing heat, and thus loss of efficiency. Loss of efficiency is minimized through the use of shorting rings on the pole piece, which, as we have discussed, has other advantages as well.

3.5 Temperature Rise and Demagnetization

With continued operation at high power input, all drivers will increase in temperature as heat develops in the voice coil and is transmitted to the surrounding parts. When magnets increase in temperature there is a loss of magnetomotive force, and this decreases the B field flux in the gap. To a large extent this is a temporary effect; when the temperature returns to normal the full flux is restored. However, Alnico is far more prone to permanent demagnetization than either ferrite (ceramic)- or neodymium based materials, so care must be taken in routine operation to avoid this.

During the heating cycle the sensitivity of the driver is diminished as a result of two effects: the loss of Bl product and the increase in the resistance of the voice coil, R_E. As a result the $(Bl)^2/R_E$ value is diminished, and there is a considerable drop in the power output of the driver for a fixed input power. As a rule, voice coil heating effects are much greater than loss of flux.

Figure 3-6 shows the degree of loss to be expected for three magnet materials

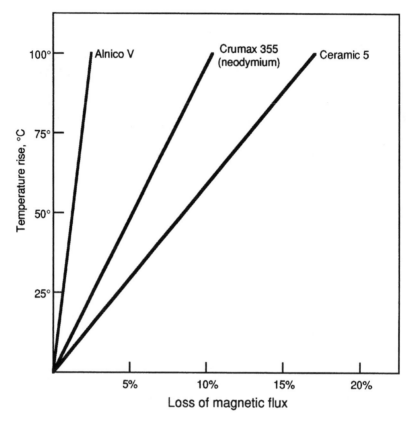

Figure 3-6. Loss of magnetic flux as a function of temperature for three magnetic materials.

as a function of temperature rise. This data, measured by Button (1992), shows that the percentage flux loss is approximately linear with regard to temperature. The examples measured here were magnetic structures in which the $H_d B_d$ product was maximized.

3.6 Modeling of Magnetic Phenomena

Not too many years ago, the design of magnetic structures was a complicated process, with a good bit of cut-and-try engineering. Today, *finite element analysis* (FEA) allows the design engineer to model the radial cross section of a magnet structure with specified magnet and iron materials and quickly get an accurate

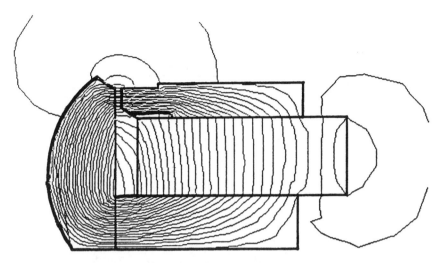

Figure 3-7. Half-section view of a magnetic structure for a compression driver. The magnet is the rectangular structure in the middle, with the top plate above. The back plate is at the bottom, and the pole piece is at the left. Flux lines are modeled via a finite element analysis program.

estimate of the final gap flux density. The effects of topology and shape can easily be seen. Figure 3-7 shows an example of such modeling (Bie, 1992). In this computer analysis, the distance between flux lines is inversely proportional to flux density, with density being the greatest through the gap region. Note also that most of the flux is contained within the structure, with only a few lines of flux (known as leakage flux) completing their return paths outside the structure. The flux paths that occur just outside the gap are known as *fringe flux*.

3.7 Magnetic Shielding

Loudspeaker systems intended for application adjacent to video monitors must be magnetically shielded in order to minimize color aberrations on the the video screen. There are several design techniques here. A bucking magnet may be attached to the back of the basic structure. In the far field, the overall stray magnetic flux will be reduced considerably, with the possible cost of 1 dB or so in loudspeaker piston band performance.

Magnetic structures that incorporate their magnets internally may not need further modification in order to achieve adequate shielding. For many computer terminals, neodymium magnets, inherently small and already internal to the overall magnetic structure, provide excellent performance in small loudspeakers.

Bibliography:

Ballou, G., ed., *Handbook for Sound Engineers,* Sams, Indianapolis, IN (1987).

Bie, D., "Design and Theory of a New Midrange Horn Driver," presented at 93rd Audio Engineering Society Convention, October 1992; Preprint No. 3429.

Buck, M., "How Woofers Work Magnetically" *Sound and Video Contractor*, Vol. 12, No. 4 (August 20, 1994).

Button, D., "Heat Dissipation and Power Compression in Loudspeakers," *J. Audio Engineering Society,* Vol. 40. Nos. 1/2 (1992).

Borwick, J., ed., *Loudspeaker and Headphone Handbook,* Butterworths, London (1988).

Collums, M., *High Performance Loudspeakers* (4th ed.), Wiley, New York (1991).

Cooke, R., ed., *Loudspeakers, Anthology,* Vols. 1 and 2, Audio Engineering Society, New York (1978 and 1984).

Olson, H., *Acoustical Engineering*, Van Nostrand, New York (1957).

Pender, H., and McIlwain, K., *Electrical Engineers' Handbook*, Wiley, New York (1950).

Sakamoto, N., *Loudspeakers and Loudspeaker Systems*, Nikankogyu Shimbun, Tokyo (1967) (in Japanese).

Low-Frequency Systems and Enclosures

4.1 Introduction

The vast majority of loudspeaker systems sold today make use of sealed or simple ported LF enclosures. The primary purpose of the enclosure is to act as a baffle, preventing the back wave of the driver from directly canceling that originating at the front of the driver. The enclosure may be sealed, in which case there is no interaction between the front and back of the driver; or it may be ported, in which case there is beneficial interaction between front and back.

In this chapter we will analyze a variety of LF enclosure types, primarily by use of the Thiele-Small (T-S) parameters. We will also briefly discuss some of the more exotic or fanciful enclosure designs that have appeared over the years.

4.2 Thiele-Small Parameters

From the earliest days of loudspeaker design to the early seventies, engineers made use of the basic electromechanical parameters of drivers in systems engineering. There was no clear-cut methodology for arriving at a given set of design goals, and the design process itself was usually an informed cut-and-try procedure.

Since the early seventies most synthesis of LF system performance has been carried out by using the T-S parameters. The work of Thiele, Small, and Benson has been widely published, and the references given at the end of this chapter will be useful, along with references to the equally important work of Locanthi and Novak. Thiele-Small parameters are routinely measured and published by driver manufacturers and provide a relatively simple method of synthesizing the high-pass filter nature of loudspeaker performance at low frequencies. Essentially, this approach is based on electrical filter theory, and it adapts many of the techniques used in that field to loudspeaker design. Much of the terminology used in the analysis is taken directly from that of filter design.

The T-S parameters are defined as follows:

f_s, free air resonance frequency of the driver's moving system.

V_{as}, equivalent volume of air that has a compliance equal to that of the driver's moving system. It is equal to $\rho_0 c^2 C_{AS}$, where C_{AS} is the acoustic compliance of the driver's suspension. Stated differently, it is the volume of air in a sealed enclosure that will raise the resonance frequency of the driver to a value of 1.4 times its free air value.

Q_{ms}, ratio of the driver's electrical equivalent frictional resistance to the reflected motional reactance at f_s.

Q_{es}, ratio of the voice coil dc resistance to the reflected motional reactance at f_s.

Q_{ts}, parallel combination of the two Q values, equal to:
$Q_{ms}Q_{es}/(Q_{ms} + Q_{es})$.

R_e, voice coil dc resistance

S_D, area of the radiating portion of the driver.

X_{max}, peak displacement capability of the moving system measured in one direction. It is nominally defined as the 10% harmonic distortion limit of the moving system.

V_D, maximum volume displacement of the cone in one direction. It is the product of X_{max} and S_D.

L_e, inductance of the voice coil.

P_E, nominal power rating of the driver, based on thermal (heating) limitations.

η_0, half-space reference efficiency, equal to $(4\pi^2/c^3)(f_s^3 V_{as}/Q_{es})$, where f_3 is the nominal -3-dB frequency and c is the speed of sound.

From these parameters, Thiele and Small derived functions that give the LF system complex response (amplitude and phase) when the enclosure volume and enclosure tuning frequency are given. They also derived functions for system impedance, cone excursion, and group delay. Measurement of the T-S parameters is discussed by Small (1972).

Today there are numerous personal computer programs available for directly plotting system performance using the above parameters, and we recommend that readers of this book familiarize themselves with one or more of these programs. As a matter of filter terminology, we refer to *alignments* of system performance when using T-S parameters. What we are aligning is, of course, the driver and the enclosure, both with their own resonance frequencies in a coupled system. Modeling of the system's response functions assumes that the enclosure is mounted in a large wall (a so-called 2π, or half-space, boundary condition).

There is a continuous range of alignments possible with any driver and enclosure, but Thiele and Small have labeled certain of the possible alignments by a convenient shorthand notation; for example, B4 stands for Butterworth, fourth order. The term *Butterworth* indicates the maximally flat passband, ripple-free cutoff response of the alignment, which is a characteristic of a Butterworth electrical filter. The term *fourth order* indicates a 24-dB/octave rolloff, each "order" contributing 6 dB to the LF rolloff. Other filter types referred to are the Chebyshev type, with equal "ripple" in the passband response, and the Quasi-Butterworth type.

The serious loudspeaker engineer will want to study the T-S parameters and their synthesis techniques in detail. Many others will be content to master one of the design programs, since these will lead to excellent designs, very nearly on a "cut-and-try" basis.

4.3 Sealed Low-frequency System Analysis

The sealed enclosure is relatively simple to analyze, since the restoring force provided by the enclosed air acts directly in parallel with the mechanical restoring force of the low frequency driver. Normally, it can be assumed that a small amount of damping material at the enclosure inner boundaries will have a relatively small effect on the alignment. In addition to damping possible standing waves in the enclosure, the use of this material has an added value in making the enclosed volume behave as though it is physically larger than it actually is. We will discuss this topic in Section 4.3.1.

Another concern with the "air spring" afforded by the sealed enclosure is its linearity. For high volume displacements the restoring force of the enclosure can be nonlinear. Normally, for volume changes that are no greater than about ±5%, the nonlinearity can be neglected.

Figure 4-1*a* shows the simulated signal output (rated power input measured at 1 m) and cone excursion of a JBL 2240 driver mounted in a sealed enclosure with a volume of 85 L (3 cu ft). In an enclosure of this size, the system response rolls off below 100 Hz at a rate approaching 12 dB per octave. Note that the excursion (scale at right) increases markedly below 200 Hz, eventually leveling off at very low frequencies just below the driver's X_{max} excursion limit. In this regard, sealed systems are "self-protecting" at high drive levels at very low frequencies. The response is simulated for rated power input and is stated in dB L_p at a distance of 1 m, as shown on the left vertical scale.

In Figure 4-1*b* we show the impedance, group delay, and phase response of this system. The free air resonance frequency of the driver is 30 Hz. When mounted in a relatively small box, as in this example, the compliance of the air spring dominates, and this raises the resonance, as shown by the impedance curve, to just below 80 Hz. The impedance curve shown in Figure 4-1*b* represents the modulus, or absolute value, of the function in ohms. The phase angle, measured in degrees, represents the deviation in driver output from the input signal. Group delay is another way of viewing this, in terms of actual signal delay. It is equal to $-d\phi/d\omega$, where ϕ is the phase angle and ω is the angular frequency, $2\pi f$. Group

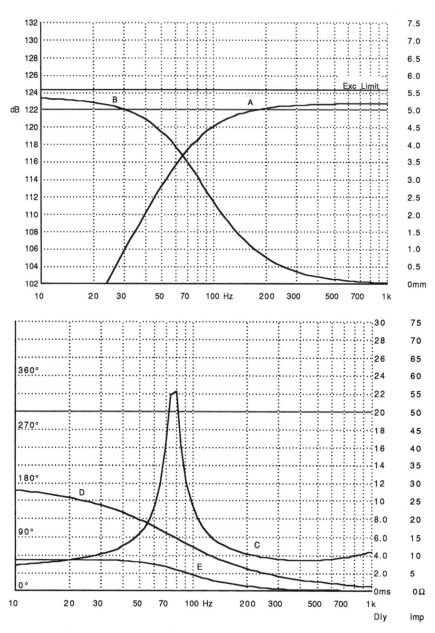

Figure 4-1. Response of a driver in a sealed system. Output amplitude (A) and cone displacement (B) at rated power (*a*); impedance (C), phase response (D), and group delay (E) (*b*).

delay is measured in milliseconds. The impedance and group delay scales are along the right axis of the graph, and the phase angle is along the left axis.

Figure 4-2 shows the effect of varying the volume of a sealed LF system, while keeping all other parameters constant. Here, we have shown the simulated response of a JBL Model 128H driver mounted in enclosures of 7, 14, 28, 56, 112, and 224 L (0.25, 0.5, 1, 2, 4, and 8 cu ft respectively). The response is quite peaked when the volume is least, progressively becoming smoother, and is finally more rolled off at low frequencies as the volume is increased. For this particular driver, note that there is little to be gained by increasing the volume beyond 56 L (2 cu ft). The system simulations are for rated power input measured at a distance of 1 m.

Figure 4-3 shows a family of curves in which the value of Q_{ts} is the only variable. This is roughly equivalent to varying the *Bl* product of the driver. Reducing *Bl* (increasing Q_{ts}) diminishes the piston band sensitivity of the system, while allowing the response at system resonance to peak progressively higher, relative to the piston band value. The driver modeled here is the JBL 128H in a 56-L enclosure, with values of Q_{ts} set at 0.24, 0.4, and 0.8 (rated power input, simulated at 1 m).

Figures 4-2 and 4-3 point out the value of the T-S parameters in analyzing

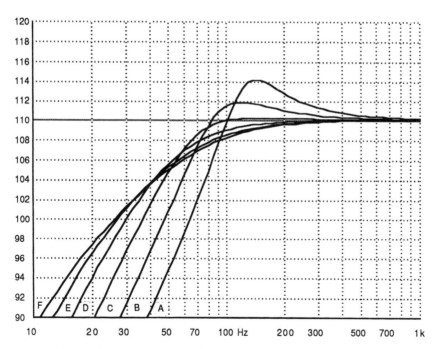

Figure 4-2. Variation in output as a function of enclosure volume: 7 L (A), 14 L (B), 28 L (C), 56 L (D); 112 L (E), and 224 L (F).

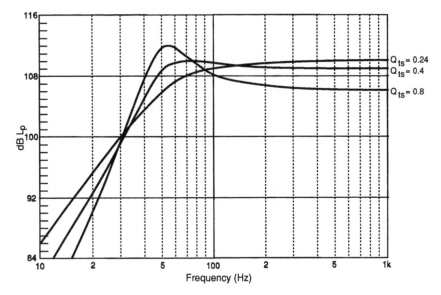

Figure 4-3. Variation in output as a function of Q_{ts}.

system performance and determining the nature of trade-offs in the design process. Figure 4-4 presents graphical data developed by Small (1972) relating the maximum efficiency of a sealed system as a function of enclosure volume and the nominal cutoff frequency (−3 dB point), f_3. A Chebyshev second order (C2) type of alignment is assumed here, with its characteristic system Q of unity.

Figure 4-5 shows the impulse response of sealed systems with varying values of Q_T at system resonance.

In sealed system design, the term *tuning ratio* α, is often used to denote the ratio of driver compliance to enclosure compliance. When the value of α is less than unity, the driver's compliance dominates. When the ratio is about 4 or higher, the compliance, or air spring, of the enclosure dominates, and the design has been referred to over the last 40 years or so as an *acoustic suspension* system. Acoustic Research, Incorporated, developed this system approach in the early fifties, virtually revolutionizing the small loudspeaker industry. See Villchur (1957) for additional information.

4.3.1 Effect of Damping Material in the Enclosure

At relatively high frequencies, standing waves may exist in the enclosure. Placing damping material on the inner walls of the enclosure will damp these out, resulting in smoother response. At low frequencies there are additional effects that derive from certain thermodynamic action. Normal acoustical processes are *adiabatic*.

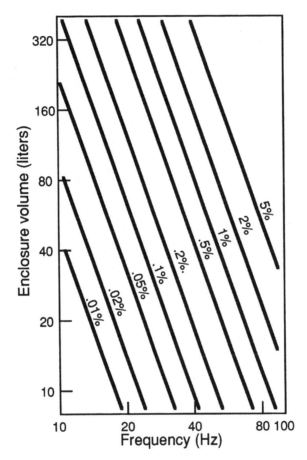

Figure 4-4. Maximum efficiency of a sealed system as a function of f_3 and enclosure volume. (Data after Small 1972.)

This implies that there is a rise and fall in the instantaneous temperature of the the air as it undergoes compressions and rarefactions in pressure.

If an enclosure is filled with damping material, the thermodynamic process becomes *isothermal*. The temperature remains relatively constant as heat is transferred to the damping material on the compression cycle, reversing itself during the rarefaction cycle. Under these conditions, the velocity of sound decreases, and this has the effect of increasing the volume of the enclosure by a significant factor. Previously it was thought that the maximum possible increase in effective volume was in the range of 1.4, or 40%, but Leach (1989) provided a more accurate analysis indicating that the maximum ratio is 1.31. In normal practice most loudspeaker engineers observe an increase of perhaps 1.2.

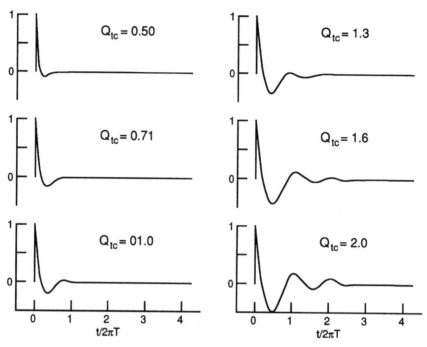

Figure 4-5. Impulse response of a sealed system as a function of Q_{tc}. (Data after Small 1972.)

The damping material should be chosen for relatively low mass and relatively high specific heat. The amount of material is usually determined empirically; too much material, tightly packed, will of course diminish the effective volume in the enclosure. If the material is placed too close to the cone, viscous losses may be significant, since the air volume velocity will be greatest close to the cone.

As a practical matter, many engineers think that the isothermal volume increase of normal amounts of damping material is roughly equal to the internal volume displaced by the driver and normal bracing in the construction of the enclosure, and they may make their initial volume calculations accordingly. In any event, the effect, to whatever degree, will manifest itself early enough in the design process to be accounted for.

4.4 Ported Low-frequency System Analysis

A ported enclosure provides a path to the outside of the enclosure, as shown in Figure 4-6a. The volume of air in the enclosure acts as an acoustical compliance, or spring, while the air in the port behaves as an acoustical mass. Sound radiates from both cone and port.

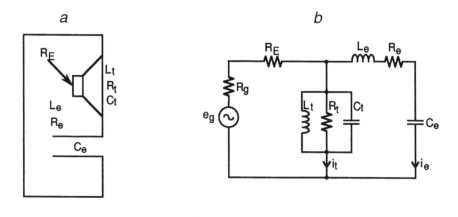

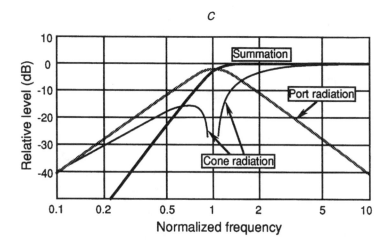

Figure 4-6. Ported systems. Section view (*a*); electrical equivalent circuit (*b*); relative output from cone and port (*c*).

The equivalent mobility electrical circuit is shown in Figure 4-6*b*, with all values reflected to the electrical input side. The acoustical output of the system is the sum of volume velocities, represented by i_t and i_e, which are the outputs of the transducer and port, respectively. In terms of acoustical output, the picture is as shown in Figure 4-6*c*. The enclosure and its port are normally tuned to a frequency in the 20–45 Hz range, representing the lower range of target system response. At that frequency the cone excursion is minimized and the volume velocity through the port is at a maximum value, producing considerable output power. The primary benefit here is that mechanical distortion in the driver is lowered because of minimized cone excursion. A secondary benefit is that it is

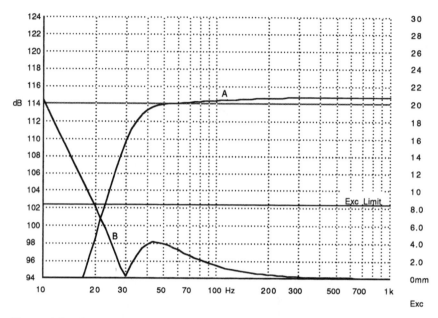

Figure 4-7. Response of a driver in a ported system. Output amplitude (A) and cone displacement (B) at rated power (*a*); impedance (C), phase response (D), and group delay (E) (*b*).

possible, with the correct driver choice, to design a LF system that can deliver substantial output power at low frequencies without the requirement of a large enclosure. The price paid for this extra measure of performance is a LF rolloff of 24 dB per octave in output below system resonance and the requirement for electrical filtering of the driving signal below resonance to avoid over-excursion of the driver. A well-thought-out alignment not only achieves the desired system response, but also takes into account system requirements in terms of thermal and displacement overload.

Figure 4-7*a* shows the total pressure output and cone excursion in a ported system. In this example, the JBL 2235H driver is mounted in a 140-L (5 cu ft) enclosure tuned to 30 Hz. The response is 3 dB down at 35 Hz, and it will be necessary to limit the signal to the system below 20 Hz if excessive cone excursion is to be minimized. The system simulation is for rated power input to the driver, measured at a distance of 1 m.

Figure 4-7*b* shows the impedance, phase response, and group delay of this system. Note that there are two peaks in the impedance response; the null in impedance between these two peaks takes place at the enclosure resonance frequency. Phase response approaches 360° at very low frequencies, while group delay is at a maximum (about 16 ms) at the impedance null. The impedance and

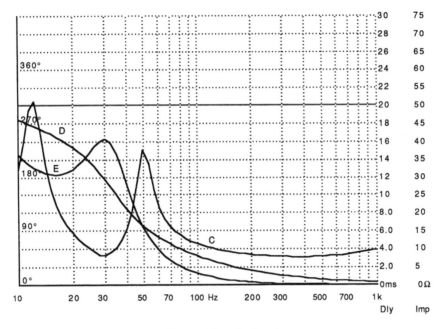

Figure 4-7. *Continued*

group delay scales are along the right axis of the the graph, and the phase angle is along the left axis.

Both sealed and ported systems have their partisans. Devotees of ported systems point out the increased LF capability for a given enclosure size, and the the devotees of sealed systems point out the more gentle LF rolloff afforded by those systems.

Ported systems involve both driver and enclosure resonances, and they present group delay and phase shift variations in the LF cutoff region in excess of that of sealed enclosures. In this regard, it is interesting to compare the data of Figure 4-1 with that of Figure 4-7. In general, it is felt that the time domain aspects of ported systems at low frequencies are swamped by the time domain nature of both music and the listening environment and may thus be ignored.

Figure 4-8 shows a useful porting chart developed by Small (1973).

4.5 Some Useful Alignments

One of the earliest benefits of the T-S analytical approach was that it allowed the loudspeaker industry to concentrate on those drivers that provided useful alignments and design them better. Today there are drivers that readily adapt to their target alignments nicely, and one rarely has to go in search of the elusive "right" driver.

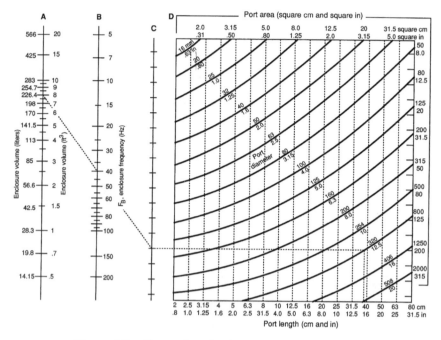

Figure 4-8. Tuning chart for determining port dimensions.

Figure 4-9 shows target Butterworth second-order (B2) and C2 alignments. The B2 alignment is maximally flat and has no ripple in the passband. In this example (Small, 1973), $f_s = 19$ Hz, $Q_{ts} = 0.32$, and $V_{as} = 540$ l (19 cu ft). Small (1972) presents the following table for determining the effect of the compliance ratio, α, on enclosure volume, system resonance, and f_3 for this example:

Table 4-1. Effect of Compliance Ratio on Enclosure Volume, Tuning, and −3-dB Frequency

α	f_c	Q_{tc}	f_3	V_{box}(L)
4	42.5	0.72	42	135
6	50.3	0.85	44	90
9	60.0	1.01	47	60
12	68.6	1.15	50	45

The C2 alignment has a Q_{tc} of unity, which causes a slight overshoot in response in the region of f_3.

One of the most useful fourth-order alignments is the Butterworth fourth-order (B4) alignment. Thiele (1971) states that this alignment is maximally flat at low frequencies when the Q_{ts} of the driver is 0.383. In the example shown here, a hypothetical driver with $f_s = 20$ Hz, $V_{as} = 815$ l (29 cu ft), and the requisite Q_{ts}.

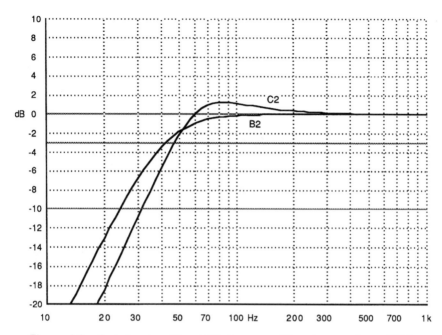

Figure 4-9. Sealed system B2 and C2 alignments. B2 enclosure volume, 137 L.

The tuning and volume ratios for the enclosure and f_3 as as given by Thiele (1971) in Table 4-2, item 1.

Table 4-2. Examples of Tuning and Volume Ratios for Fourth-Order Alignments

	Type	f_3/f_s	f_3/f_b	V_{as}/V_{enc}	Q_t
1.	B4	1.00	1.00	1.414	0.383
2.	C4	.641	0.847	0.559	0.518

Note that for the B4 system, the enclosure tuning frequency is set equal to f_s and that the ratio of V_{as} to enclosure volume (V_{enc}) is 1.414. Using these values gives the simulated response shown in Figure 4-10a.

This example points out very clearly the advantages of T-S analysis and drivers that have been engineered for specific alignments. The extended LF response exhibited here (-3 dB at 23 Hz) would be difficult to achieve with most high-powered professional LF drivers designed for high piston band sensitivity, with their lower values of Q_{ts}. Thiele and Small have shown that a less efficient (and less expensive) driver will actually perform better in the LF range.

Continuing in the same vein, the Chebyshev fourth-order (C4) alignment provides further LF extension, using an even lower efficiency driver with

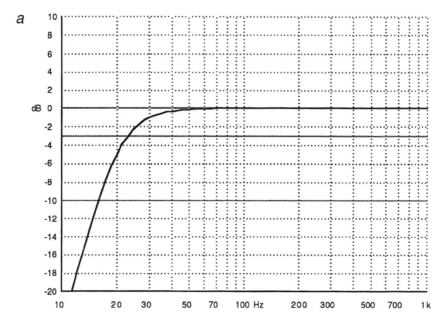

Figure 4-10. Ported system alignments. B4 alignment (*a*); C4 alignment: curve A, without enclosure losses; curve B, with enclosure losses (*b*).

$Q_{ts} = 0.518$. Tuning and volume ratios are given as item 2 in Table 4.2, and the response is shown as curve A in Figure 4-10*b*. Curve A assumes that the enclosure has negligible losses; more likely, we would see a curve such as B in modeling this alignment with a typically "lossy" box. In either case, the LF extension down to about 0.707 of the driver's resonance is exemplary. Such a system would require very careful monitoring of its input signal in the subsonic range.

Another very useful alignment is the so-called Butterworth type, sixth-order (B6) first discussed by Thiele (1971) and later by Keele (1975). In this alignment, a ported system is first designed to exhibit flat response (B4 alignment). The enclosure tuning is then reduced by a factor of 0.707 to produce an overdamped low frequency response, and is subsequently equalized for flat acoustical output in the bass region by using a 6-dB electrical boost with a Q of 2. If carefully done, the design exhibits smooth response over its passband and maintains good immunity to displacement overload at full thermal rating. Figure 4-11 shows the rolled-off natural response, the electrical filter response, and the net acoustical output.

ElectroVoice has designed a number of LF systems that make use of what they term a "step-down" mode. Normally, the system is of the B4 type, and can be converted to a B6 alignment by reducing the port area with an appropriate cover and equalizing the system accordingly. The step-down kit provides these elements.

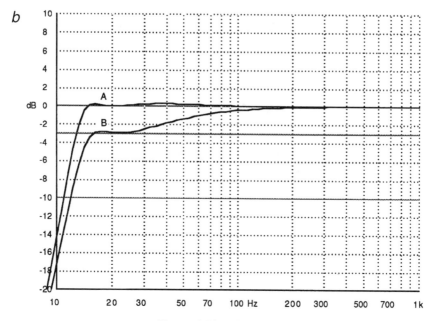

Figure 4-10. *Continued*

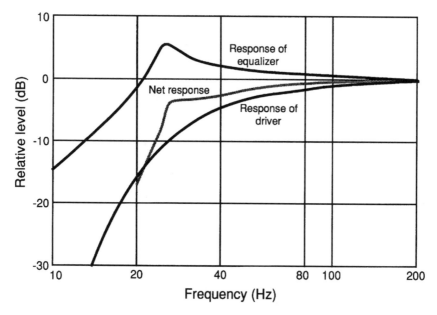

Figure 4-11. The Butterworth type, sixth-order (B6) alignment.

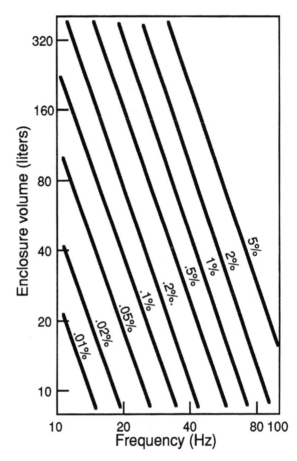

Figure 4-12. Maximum efficiency of a ported system as a function of f_3 and enclosure volume. (Data after Small 1973.)

Figure 4-12 shows the relation between enclosure volume, f_3, and system efficiency that can be realized with ported systems. Figure 4-13 shows the response of ported systems to an impulse function as a function of Q_{tc}. It will be instructive to compare these two figures with their corresponding ones for sealed systems, Figures 4-4 and 4-5.

4.6 The Passive Radiator

Figure 4-14*a* shows a section view of a LF system with a passive radiator ("drone cone" and auxiliary bass radiator are other terms for the passive radiator). The passive radiator (PR) resembles a cone driver without a motor; it is effectively a mass with related compliance and mechanical damping used in place of a port.

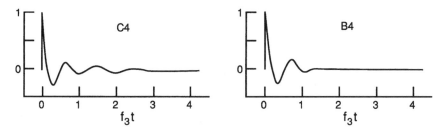

Figure 4-13. Impulse response of a ported system as a function of Q_{tc}. (Data after Small 1973.)

The PR is normally chosen to be of the same diameter as the active driver. As shown in Figure 4-14*b*, the simplified mobility circuit shows the compliance of the PR in parallel with its mass. It is the added compliance that sets the system apart from the standard ported system, and if the compliance is made large enough, the behavior of the system over the useful passband is very close to that of a conventional ported system (Small 1974).

The benefit of a PR is primarily the reduction of port air turbulence that may afflict a ported system if its port area is inadequate. On the debit side, the PR normally suffers from the nonlinearity of its suspension elements at high excursions. Once in vogue, the PR is now generally felt to be a thing of the past.

4.7 Transmission Line Systems

There are several types of enclosures that qualify for the term "transmission line." Basically, these are enclosures that are long enough in one internal dimension to accommodate at least one-quarter wavelength at the lowest design frequency of the system. The path is usually folded and lined with damping material, as shown

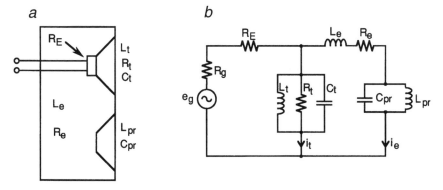

Figure 4-14. The passive radiator. Section view of system (*a*); electrical equivalent circuit (*b*).

a b

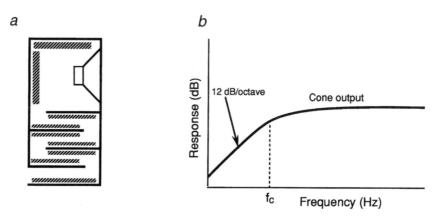

Figure 4-15. Section view of a transmission line system with dissipative duct.

in Figure 4-15*a*. The system may or may not be open at the end of the transmission line. In this application the line acts as an infinite baffle, but one that does not add compliance in series with the driver and thus raising the system's resonance frequency. The effect of the damping material is to make the line lossy so that reflections from the end of the line back toward the driving driver are attenuated.

Details of the Jensen Transflex system are shown in Figure 4-16. Here, the transmission line is one wavelength long at the lowest system frequency, and it is folded back on itself so that the "organ pipe" resonance provides considerable reinforcement at the design frequency. These systems were once build under living room floors, where the structural members became integral parts of the enclosure. Tuning frequencies were normally in the range of 25–35 Hz, and the effective system response extended to about one octave. An excellent analysis of the technology is given by Tappan (1959).

4.8 Curiosities from the Past

During the 1940s and 1950s many exotic loudspeaker enclosures were developed, only to disappear a few years later. At best, many of these systems did improve some aspect of the performance of inadequate drivers.

The R-J enclosure, as shown in Figure 4-17*a*, provided a direct path from the full-range driver to the outside. At low frequencies the exit from the enclosure was augmented by a narrow distributed slot that added air mass and viscous losses to the cone's output. It is easy to see how such an enclosure could damp and possibly extend the LF response of a small driver with a light cone and relatively small motor.

Figure 4-17*b* shows details of the Karlson enclosure. According to the designer's description, the folded back-path terminated on the front with a tapered slot

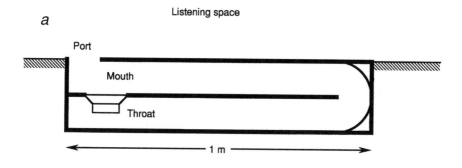

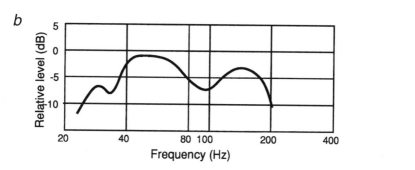

Figure 4-16. Section view of Jensen Transflex transmission line system (*a*); typical response (*b*). (Data after Tappan, 1959.)

that "detuned" the normal resonance of the back-path (assumed to be long enough to be a transmission line), making it responsive over a broad band—hardly an accurate description. The enclosure was further recommended for use with full-range drivers, and one can imagine the the midrange resonances that the front cavity caused.

The designer of the Bradford Perfect Baffle, shown in Figure 4-17c, apparently thought that mounting a small vertically hanging flap in a hole on the rear of the enclosure would rid the enclosure of back pressure from the driver and make it perform like a truly infinite baffle! A better description might be "imperfect passive radiator."

4.9 Multichamber Bandpass Low-frequency Systems

In the normal ported enclosure there is useful output from both cone and port. The high pass nature of the system is governed by the driver-enclosure alignment,

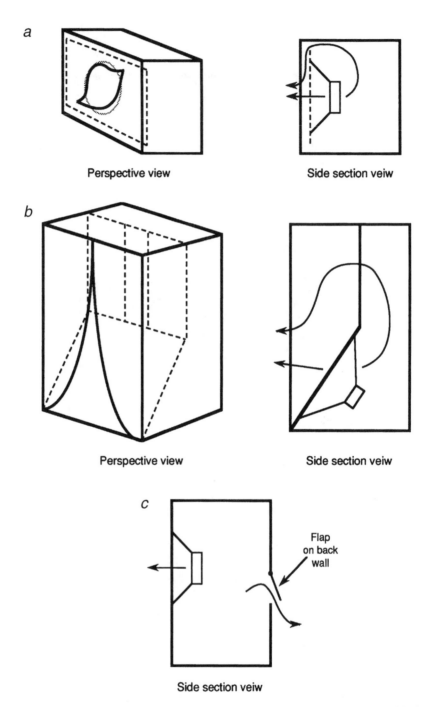

Perspective view

Side section veiw

Perspective view

Side section veiw

Flap on back wall

Side section veiw

Figure 4-17. Unusual enclosures. The R-J enclosure (*a*); the Karlson enclosure (*b*); the Bradford Perfect Baffle (*c*).

while the low-pass nature of the system is governed by the diameter of the cone and related values of *ka*.

In multichamber LF systems, the driver is physically located between two of the chambers. Two or more chambers may be used, and both low- and high-pass response of the system is governed by tunings of the enclosures. The advantages of these systems are bandpass response without the need for electrical equalization and, in some cases, lower overall chargeable volume for their bandpass and output capabilities. On the debit side, these systems are physically complex and are relatively susceptible to boundary (mounting) conditions.

The form shown in Figure 4-18*a* consists of a driver in a sealed chamber. A second chamber is placed in series with the driver and is ported to the outside. The equivalent circuit is shown at *b*, and the nominal response is shown at *c*. In this design, f_1, the lower cutoff frequency, is determined primarily by the inner chamber. f_2, the upper cutoff frequency, is determined by the outer chamber. This design is in the public domain.

The form shown in Figure 4-19*a* consists of a second ported chamber in series with both the cone and port outputs of a normally ported inner enclosure. The equivalent circuit is shown at *b*, and nominal response is shown at *c*. This design, at least in the United States, is covered by patents held by Teledyne and Bose, the distinctions being based on relative volumes of the two chambers and and their tuning ratios.

The form shown in Figure 4-20*a* consists of parallel tuned chambers on either

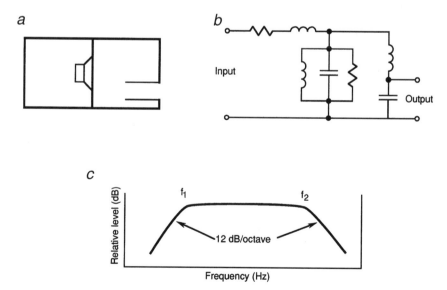

Figure 4-18. Bandpass system: sealed-series operation. Section view (*a*); equivalent circuit (*b*); target response (*c*).

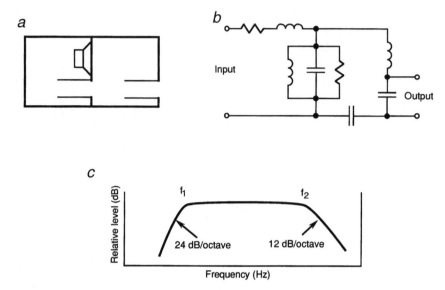

Figure 4-19.   Bandpass system: ported-series operation. Section view (*a*); equivalent circuit (*b*); target response (*c*).

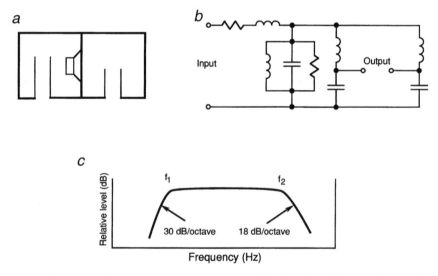

Figure 4-20.   Bandpass system: parallel operation. Section view (*a*); equivalent circuit (*b*); target response (*c*).

side of a driver. The equivalent circuit is shown at *b* and nominal response is shown at *c*. In the United States, this design is covered by a patent held by Bose. Elsewhere, prior art, most notably a French patent issued to d'Alton in 1937, governs.

In all of these designs the ratio of f_1 and f_2 are critical in the proper functioning of the systems. If these frequencies are too closely spaced, the system coalesces into a single resonance; if too far apart the ideal bandpass is compromised.

4.10. Transducers in Acoustical Series and in Parallel

4.10.1 Parallel Operation

It is common in many applications to mount two or more LF drivers adjacent to each other on the same baffle, normally driving them electrically in parallel. At the same time, the drivers are acting acoustically in parallel. Figure 4-21*a* shows a front view of a LF system consisting of two JBL Model 2226 380-mm (15-in.) drivers mounted in a 225-L enclosure tuned to 40 Hz. The curves shown in Figure 4-21*b* show the power response of this system, as compared to that of a single 2226 driver mounted in a 112-L enclosure tuned to 40 Hz. In both cases, 1 w of power has been applied to each system.

Note that the response of the dual driver system is 3 dB greater than the single unit, both with the same electrical power input. This increase is caused by the fact that the two closely coupled drivers behave essentially like a single "new" driver with twice the cone area, twice the V_{as}, and the same f_s and Q_{ts}. The doubling of cone area will result in a doubling of efficiency, but the overall LF alignment will remain essentially the same. The HF rolloff commencing at $ka = 2$ will be shifted downward by a factor of 0.7 in frequency, since the effective

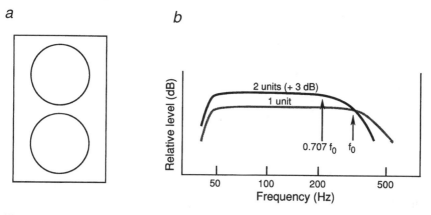

Figure 4-21. Low-frequency drivers in acoustical parallel. System front view (*a*); response, as compared to a single driver (*b*).

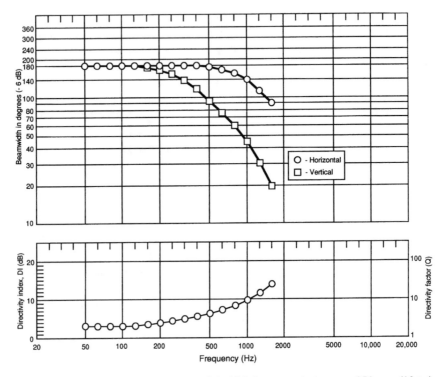

Figure 4-22. Directional properties of dual LF driver vertical arrays. 250-mm (10-in.) drivers (*a*); 300-mm (12-in.) drivers (*b*); 380-mm (15-in.) drivers (*c*).

perimeter of the radiating surface has been increased by a factor of about 1.4. The term *mutual coupling* is often used to describe this effect.

It is further obvious that the dual driver system can handle twice the electrical input power than the single unit. Therefore, there is a net 6 dB greater output capability with the dual unit as opposed to the single unit.

4.10.2 Directional Properties

The nominal −6 dB beamwidth and DI of three sizes of dual LF systems are shown in Figure 4-22*a–c*. The assumption is made that the drivers are arrayed in a vertical line, so that maximum coverage will be obtained horizontally.

4.10.3 Series Operation

The arrangement shown in Figure 4-23 places a pair of drivers one behind the other with a small enclosed airspace between them. The pair is normally driven electrically in parallel, but they are acting acoustically in series. Colloms (1991) describes the combination as acting like a single driver having twice the moving

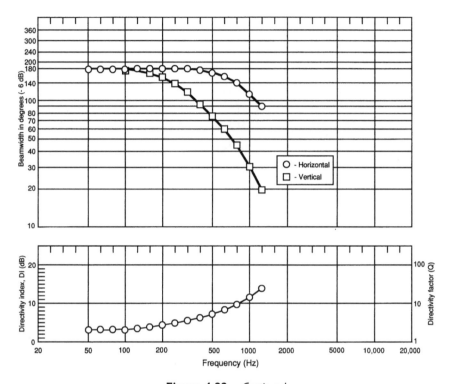

Figure 4-22. *Continued*

mass, half the compliance, and half the impedance of the single driver. For a given voltage input, it will absorb twice the power of the single unit, and in a sealed enclosure, the air spring will be the dominant restoring force. The net result of this will be a reduction of the system resonance to 0.7 that of the single driver in the same enclosure, with an increase in power handling and linearity.

Olson (1957) presents a detailed analytical description of the series driver arrangement, essentially showing that LF response is extended to 0.7 the cutoff frequency for a single driver mounted in an enclosure of the same volume.

4.11 Alignment Shifts

Large arrays of "subwoofers," such as may be used in high level music reinforcement or in motion picture theaters, exhibit a shift in alignment, as shown in Figure 4-24. Here, we show the shift in the impedance null at enclosure resonance observed with one, two, and four ported LF systems. Measurements were made with the systems placed on an extended ground plane outdoors.

The progressive shift of the impedance null from 31 Hz, downward to 29 and 27 Hz, indicates a trend that, if extended much beyond the point shown, could

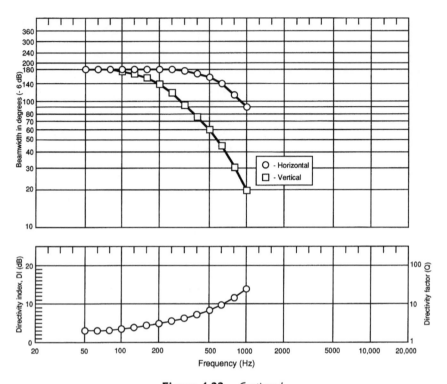

Figure 4-22. *Continued*

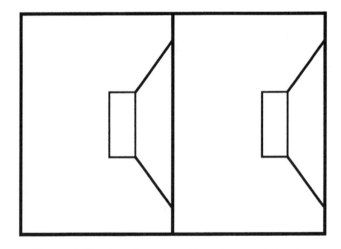

Figure 4-23. Low-frequency drivers in acoustical series.

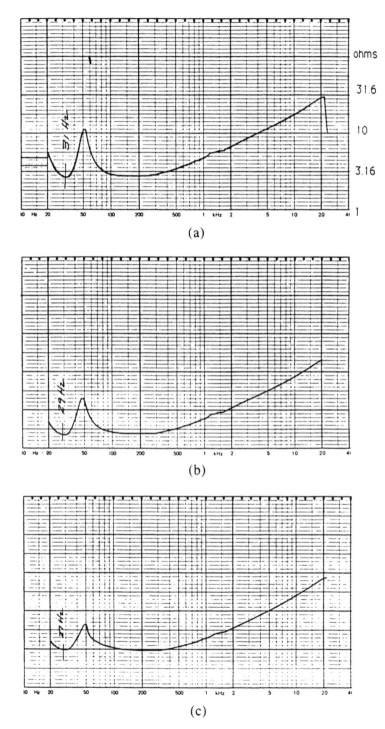

(a)

(b)

(c)

Figure 4-24. Alignment shifts in multiple LF system operation. Single system (a); two systems (b); four systems (c). (Data courtesy J. *Audio Engineering Society*; Gander and Eargle, 1990.)

83

compromise system performance. The downward shift may make the combined system of LF units susceptible to displacement overload in the octave above enclosure resonance and should be carefully monitored in this regard. In extreme cases it may be necessary to realign the LF systems, taking these shifts into consideration.

Bibliography

Augspurger, G., "Theory, Ingenuity, and Wishful Wizardry in Loudspeaker Design—a Half-Century of Progress?" *J. Acoustical Society of America*, Vol. 77, No. 4 (1985).

Avedon, R., "More on the Air Spring and the Ultra-Compact Loudspeaker," *Audio Magazine*, June (1960).

Beranek, L., *Acoustics*, McGraw-Hill, New York (1954).

Borwick, J., (ed.), *Loudspeaker and Headphone Handbook*, Butterworth, London (1988)

Colloms, M., *High Performance Loudspeakers*, 4th ed., Wiley, New York (1991).

Cooke, R., ed., *Loudspeakers, Anthology,* Vols. 1 and 2, Audio Engineering Society, New York (1978 and 1984).

Engebretson, M., "Low-Frequency Sound Reproduction," *J. Audio Engineering Society*, Vol. 32, No. 5 (1984).

Gander M., and Eargle, J., "Measurement and Estimation of Large Loudspeaker Array Performance," *J. Audio Engineering Society*, Vol. 38, No. 4 (1990).

Geddes, E., "An Introduction to Band-Pass Loudspeaker Systems," *J. Audio Engineering Society*, Vol. 37, No. 5 (1989).

Keele, D., "A New Set of Sixth-Order Vented-Box Loudspeaker System Alignments," *J. Audio Engineering Society*, Vol. 23, No. 5 (1975).

Leach, M., "Electroacoustic-Analogous Circuit Models for Filled Enclosures," *J. Audio Engineering Society*, Vol. 37, No. 7/8 (1989).

Locanthi, B., "Application of Electric Circuit Analogies to Loudspeaker Design Problems," *J. Audio Engineering Society*, Vol. 19, No. 9 (1971).

Novak, J., "Performance of Enclosures for Low-Resonance High-Compliance Loudspeakers, *J. Audio Engineering Society*, Vol. 7, No. 1 (1959).

Olson, H., *Acoustical Engineering*, Van Nostrand, New York (1957).

Small, R., "Direct Radiator Loudspeaker System Analysis and Synthesis (parts 1 and 2)," *J. Audio Engineering Society*, Vol. 20, No. 5 (1972) and Vol. 21, No. 1 (1973).

Small, R., "Passive-Radiator Loudspeaker Systems Part I: Analysis," *J. Audio Engineering Society*, Vol. 22, No. 8 (1974).

Tappan, P., "Analysis of a Low-Frequency Loudspeaker System," *J. Audio Engineering Society*, Vol. 7, No. 1 (1959).

Thiele, A. N., "Loudspeakers in Vented Boxes, Parts I and II," *J. Audio Engineering Society*, Vol. 19, Nos. 5/6 (1971).

Villchur, E., "Problems of Bass Reproduction in Loudspeakers," *J. Audio Engineering Society*, Vol. 5, No. 3 (1957).

Dividing Networks and Systems Concepts

5.1 Introduction

Loudspeaker engineers may generally agree on what is or is not a good loudspeaker, but they may argue at length on the relative importance of the attributes that define a good loudspeaker. Often, the refinement of one performance attribute comes at the expense of another, and there are economic trade-offs in the world of commerce. There are many design variables that must be dealt with, and in this chapter we will cover systems concepts, details of dividing network topology and design, off-axis lobing, or interference, problems, baffle layout and enclosure edge treatment.

We also stress that there are three loudspeaker attributes that are difficult to reconcile: enclosure size (smaller is better), bandwidth (extended LF response is better), and efficiency (higher is better). This has often been referred to as the "eternal triangle" of loudspeaker design. Any improvement in one of these attributes always comes at the expense of the other two, and the design of commercial loudspeaker systems involves continuous trade-offs among these attributes.

5.2 Basic Dividing Networks

The function of a frequency dividing network is to channel the audio spectrum into two or more bands so that the system drivers receive their intended signals. The necessity for this is the fact that a loudspeaker system is a collection of bandpass devices, each requiring its share, and only its share, of the input spectrum. Additionally, it is imperative to keep LF signals from reaching HF drivers and possibly damaging them.

Networks are primarily defined in terms of their nominal rolloff slopes. As discussed in Chapter 4, the *order* of a network slope defines the number of reactive elements contributing to the rolloff, each element producing a slope of

6 dB/octave. Thus, a first-order network produces a 6 dB/octave slope, while a fourth-order network will produce a 24 dB/octave slope. The term *pole* is often used to describe the network action that contributes a 6 dB/octave rolloff.

Figure 5-1 shows design data for first- and second-order two-way networks in both parallel and series form. The first-order networks are shown at *a* and *b*, and both make use of the same circuit values for a given crossover frequency. The parallel form is used more often, but an advantage of the series form is that the network slopes in the region of crossover are very nearly second-order, becoming first order at frequencies removed from the crossover point.

Second-order networks are shown at *c* and *d*, with their circuit values expressed in terms of L_1 and C_1. Note that the circuit values are not the same for parallel and series forms.

In using the network types described in Figures 5-1, it is essential that the design load impedances closely match the actual impedance values of their respective drivers at the selected crossover frequencies. All target values should be constructed and measured to ensure that they are correct.

A three-way parallel second order network is shown in Figure 5-1*e*. Here, the MF section is bandpass, composed of both LF and HF sections. Circuit values are obtained using the data given in Figure 5-1.

5.2.1 Vector Relationships in Networks

Figure 5-2*a* shows the vector relationship at crossover between high- and low-pass sections for a first-order network. Note that the amplitude of both high and low signals at crossover is 0.7 and that both are displaced ±45°. The on-axis summation is unity at a phase angle of zero, and the power response, proportional to $\sqrt{(0.7)^2 + (0.7)^2}$, is likewise unity. These are both desirable attributes, but the chief problem with first-order networks is that the out-of-band signals may not be sufficiently attenuated for proper component protection, depending on the system's application.

Second-order vector relationships are shown in Figure 5-2*b*. Here the two signals are each shifted 90° for a total of 180°. In this case it is necessary for one of the two outputs of the network to be reversed in polarity so that the signals will add at the crossover frequency.

5.2.2 Higher Order Networks

Third order networks are often used in professional applications, where considerable out of band attenuation is necessary because of the high powers that may be involved. A typical circuit is shown in Figure 5-3*a*, and the normal vector relationships are as shown in Figure 5-3*b*. If the HF output of the network is inverted the system phase response will change, as shown in Figure 5-3*c*.

The so-called Linkwitz-Riley network is of fourth-order and is normally de-

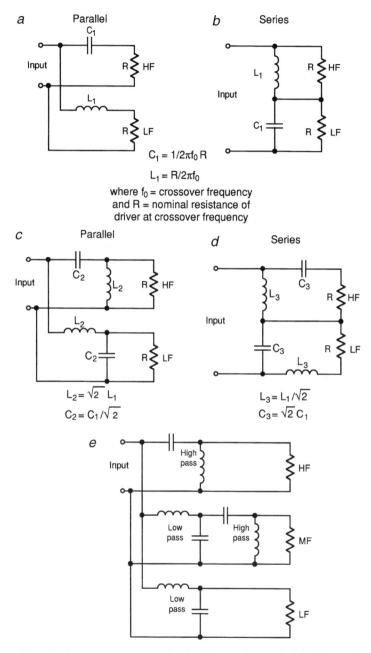

Figure 5-1. Basic two-way network circuits. First-order parallel (*a*); first-order series (*b*); second-order parallel (*c*); second-order series (*d*); second-order three-way parallel network (*e*).

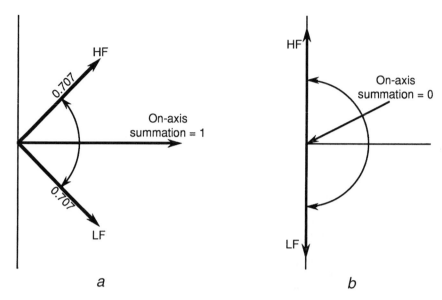

Figure 5-2. Vector relationships at the crossover frequency. First-order (*a*); second-order (*b*).

signed so that both HF and LF outputs are 0.5 amplitude (−6 dB) at the crossover frequency. Both sections are rotated 180° at crossover and will of course add to produce unity output on axis, as shown in Figure 5-4*a*. Power response will dip 3 dB at crossover, however, as shown at *b*. The advantages of the design is that out-of-band signals are greatly attenuated and that lobing effects between adjacent drivers will take place over a very small frequency interval.

The network circuits we have discussed thus far can be cascaded to provide three- and four-way designs, as required. However, parallel networks may be easier to work with in the design of three-way or higher systems.

5.2.3 Conjugate Networks

Conjugate networks can be used to compensate for the effects of resonance in drivers, as well as the impedance rise at HF due to driver inductance. An example of this is shown in Figure 5-5. The basic driver is shown in Figure 5-5*a* and its impedance in Figure 5-5*b*. The addition of an impedance compensation network (shown in Figure 5-5*c*) will alter the impedance, as shown in Figure 5-5*d*. Further addition of an inductance compensating conjugate network (shown in Figure 5-5*e*) will further flatten the impedance curve, as shown in Figure 5-5*f*.

When a driver has been compensated as shown in Figure 5-5 it may effectively be treated as a resistance, and network values calculated in a straightforward manner. The resulting crossover slopes will be well behaved, since there will be

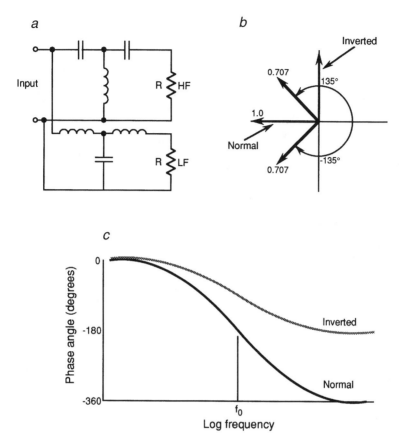

Figure 5-3. Third-order network design. Parallel circuit (*a*); vector relationships (*b*); phase response, normal and inverted (*c*).

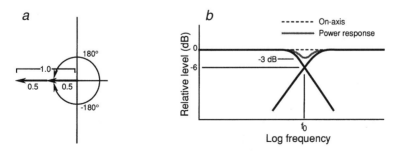

Figure 5-4. Fourth-order design (Linkwitz-Riley). Vector relationship at crossover (*a*); system response (*b*).

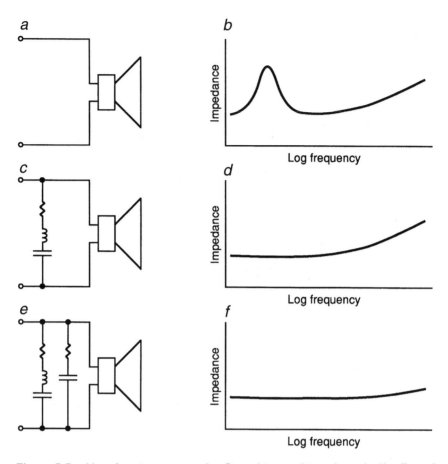

Figure 5-5. Use of conjugate networks. Cone driver and impedance (*a*, *b*); effect of resonance conjugate network (*c*, *d*); effect of inductance conjugate network (*e*, *f*).

very little reactive loading on the dividing network. The calculation of circuit values for impedance compensation is provided by a number of the system design programs that are available. For the voice coil inductance compensating network, the value of capacitance, in farads, is given by:

$$C = L/(Z)^2 \tag{5.1}$$

where L is the voice coil inductance and Z is the nominal impedance of the driver. The value of R in the inductance compensating network is the design impedance of the driver.

The resonance compensating network component values are given by

$$L = Q_t R/2\pi f_0 \tag{5.2}$$

where Q_t is the system Q at resonance, R is the nominal impedance of the driver, and f_0 is the resonance frequency. L is given in henrys.

$$C = 1/L(2\pi f_0)^2 \qquad (5.3)$$

where C is given in farads and L in henrys.

5.2.4 Upper Useful Frequency Limits for Drivers

As we have discussed earlier, cone-type loudspeakers begin to roll off in power response at values of ka of 2 and higher. However, due to the accompanying increase in their on-axis directivity index, it is possible to use these drivers at values higher than $ka = 2$. Table 5.1 presents useful upper-frequency limits, based on DI values of 6 and 10 dB. As a general rule, taking a cone driver higher than a DI of 10 is considered a bit risky, in that manufacturing tolerances at high frequencies may come into play, and unit to unit consistency may be questioned.

Table 5.1   Useful Upper Frequency Limits for Cone Drivers

Diameter	$ka = 2$ (DI = 6 dB)	DI = 10 dB
460 mm (18 in.)	547 Hz	820 Hz
380 mm (15 in.)	673 Hz	1010 Hz
300 mm (12 in.)	875 Hz	1313 Hz
250 mm (10 in.)	1100 Hz	1650 Hz
200 mm (8 in.)	1460 Hz	2190 Hz

5.2.5 Some Notes on Network Component Quality

High-quality network components should be used whenever possible, and careful note should be taken of working voltages so that the components will not be stressed in normal use. Inductances in particular need to be made of fairly large-gauge wire so that their associated resistance can be minimized. Iron core inductors may be used, but only if effects of core saturation, and the consequent shift in inductance, are minimal and accounted for in the design calculations. Powdered core inductors are less prone to inductance shifts.

When higher-sensitivity drivers are padded down to match lower-sensitivity drivers, the padding resistors may dissipate a good bit of heat when the system is heavily driven. Choose the sizes carefully. In particular, watch the gauge of rotary "L-pads" that are often used for adjustable driver levels on many systems.

Many of the capacitors used in dividing networks are fairly large and require substantial working voltages. Nonpolarized electrolytic types may be used in shunt network paths, but only the highest-quality components should be specified for primary signal paths. Many engineers routinely use high-quality, small-value bypass capacitors placed in parallel around the larger ones, in an effort to linearize the overall response. The jury is still out regarding this technology.

5.3 Stock Networks and Autotransformers

For many professional applications, where there may be mixing and matching of separate HF and LF systems, stock networks are often used. Figure 5-6a shows a photograph of typical examples. A stock network schematic is shown in Figure 5-6b and typical response curves are shown in Figure 5-6c. The basic design is as a second-order network with variable HF output so that the HF driver sensitivity (normally a horn system) can be properly matched to the LF section. A switch allows HF power response boost equalization to be introduced into the HF output, with typical curves shown in Figure 5-6d.

Such networks are normally designed as a "best fit" for a number of applications and may not represent an ideal crossover solution to any one of them.

Details of the autotransformer (also known as autoformer) are shown in Figure 5-7a. This single winding version of the transformer can be used to adjust impedances and drive levels in a network as shown. In the application shown in Figure 5-7b, a 4-Ω load is transformed so that it appears as a 16-Ω load. When a load impedance is transformed to a higher value, it will absorb less power. Accordingly, the autoformer is often used to adjust drive levels at the output of passive network designs.

5.4 Combining Acoustical and Electrical Poles

The natural rolloff of a driver, if it is smooth and controlled, can be used as a network pole, or poles, contributing to the effective order of the network. In the example shown in Figure 5-8 the LF driver has a natural rolloff of 12 dB/octave commencing at about 1 kHz and sensitivity of 90 dB, 1 W at 1 m. If a second-order low-pass section is added to the driver the net output slope above 1 kHz will be of fourth order.

Figure 5-9 shows the application of a first-order high-pass section, adding a single acoustical pole to the network action. The HF driver under consideration here exhibits a natural third-order high-pass action, so only a single capacitor can be used to arrive at a net fourth-order transition. In this case, the midband sensitivity of the driver-horn combination is about 114 dB, 1 W at 1 m, around which a HF bypass network allows the overall response to be extended to just beyond 10 kHz. This equalization is possible, inasmuch as the HF section has excess midband sensitivity which can be sacrificed for a better match with the LF unit.

The combined on-axis acoustical action of the two sections is shown in Figure 5-10, and the dividing network schematic circuit is shown in Figure 5-11.

5.5 Off-Axis Lobing Effects

While the response of combined HF and LF radiators can be made flat on-axis, the response observed off-axis in the the plane of the two drivers may exhibit

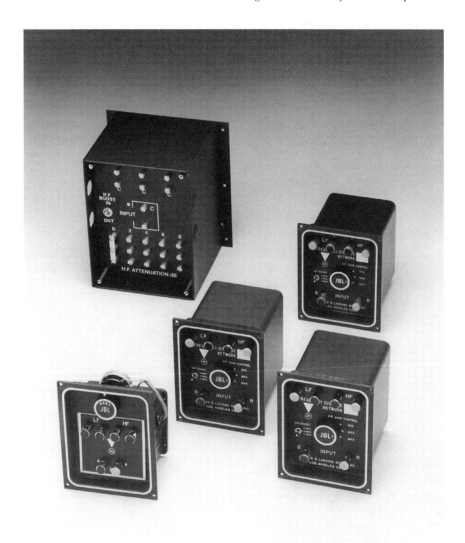

a

Figure 5-6. Stock networks. Photograph (*a*); circuit diagram (*b*); normal response range (*c*); network response with HF power response correction (*d*). (Data courtesy JBL Inc.)

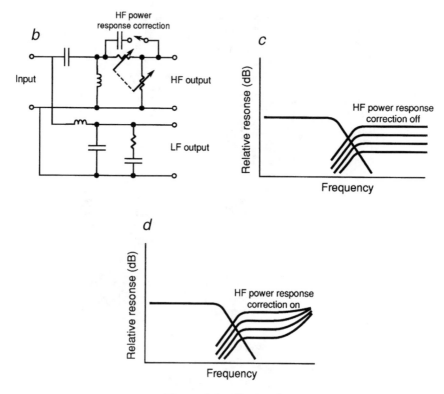

Figure 5-6. *Continued*

lobing errors. The less lobing error a system exhibits, the less critical the listening angle will be; the greater the lobing error, the more restricted the listening angle will be.

In the following examples of lobing error, we are assuming that the HF and LF components are vertically mounted in an enclosure and that the effective distance between them is 40 cm (16 in.), as shown in Figure 5-12a. A crossover frequency of 1 kHz is assumed.

For a second-order network, as shown in Figure 5-2b, normal connection of the HF device will exhibit a null on axis. This is shown as a lobing error in the vertical plane, as shown in Figure 5-12a and b. Reconnecting the HF element in reverse polarity will result in the lobing response shown in Figure 5-13. With this corrected condition, the lobing response half an octave above crossover (1414 Hz) will be as shown in Figure 5-14, and at half an octave below crossover (707 Hz) the lobing response is as shown in Figure 5-15.

Figure 5-16 shows the lobing response of a third-order network in which the HF element has been connected in reverse polarity, as shown in Figure 5-3c.

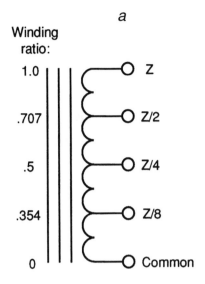

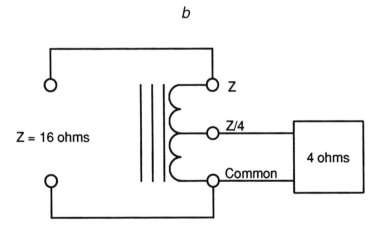

Figure 5-7. The autoformer. Circuit (*a*); typical application (*b*).

Here, the major lobe is skewed slightly downward. For this same connection, the lobing response half an octave above crossover (1414 Hz) will be as shown in Figure 5-17, and the lobing response half an octave below crossover will be as shown in Figure 5-18.

The Linkwitz-Riley fourth-order network exhibits a lobing response on-axis

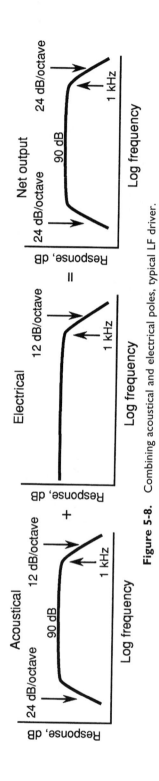

Figure 5-8. Combining acoustical and electrical poles, typical LF driver.

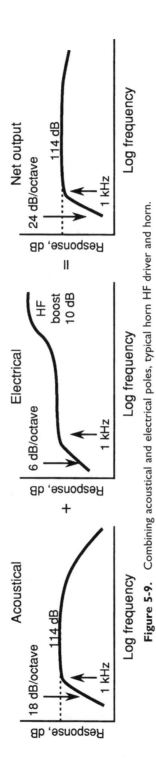

Figure 5-9. Combining acoustical and electrical poles, typical horn HF driver and horn.

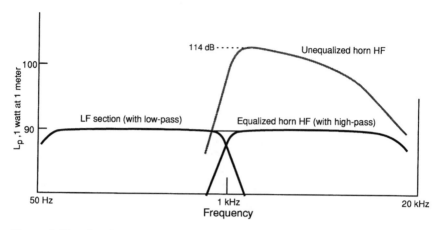

Figure 5-10. Combining acoustical and electrical poles, combined HF and LF response.

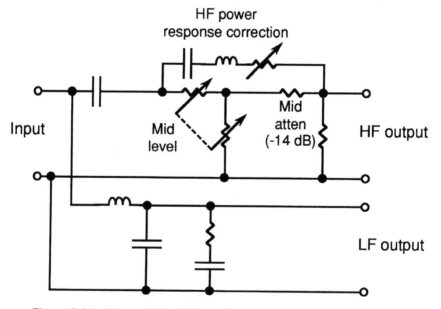

Figure 5-11. Network details for combining acoustical and electrical poles.

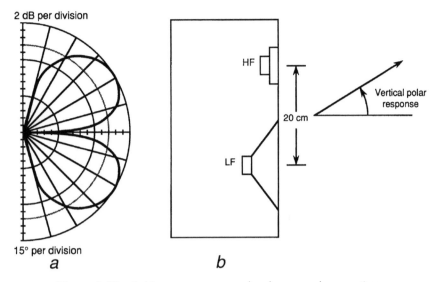

Figure 5-12. Lobing response, second-order, normal connection.

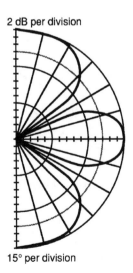

Figure 5-13. Lobing response, second-order, reversed connection.

2 dB per division

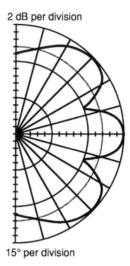

15° per division

Figure 5-14. Lobing response, second-order, half an octave above crossover.

2 dB per division

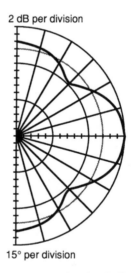

15° per division

Figure 5-15. Lobing response, second-order, half an octave below crossover.

as shown in Figure 5-19. Figures 5-20 and 5-21, respectively, show the lobing response at half an octave above crossover (1414 Hz) and at half an octave below crossover (707 Hz).

The main observation to be made here is that lobing errors are minimized when the wavelength at the crossover frequency is large relative to the spacing of adjacent HF and LF drivers. A concern for this situation normally leads

2 dB per division

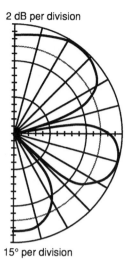

15° per division

Figure 5-16. Lobing response, third-order, reversed connection.

2 dB per division

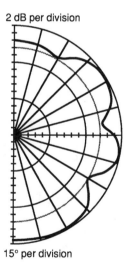

15° per division

Figure 5-17. Lobing response, third-order, half an octave above crossover.

loudspeaker engineers to specify very small distances between elements whenever possible.

5.6 Baffle Component Layout and Edge Details

Figure 5-22 shows a variety of front views of typical baffle layouts. A symmetrical vertical array is shown at *a*. This is typical of most loudspeakers on the market

2 dB per division

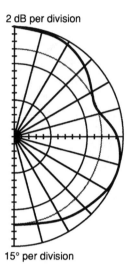

15° per division

Figure 5-18. Lobing response, third-order, half an octave below crossover.

2 dB per division

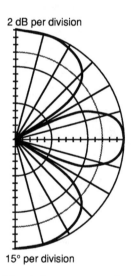

15° per division

Figure 5-19. Lobing response, fourth-order.

today in that the design does not require separate left and right models. Due to cost constraints, most rectangular enclosures are made with no rounding of their edges, and this aggravates certain boundary conditions.

The design shown in Figure 5-22b does not have left-right symmetry, and it is necessary for separate left and right models to be offered as a pair.

2 dB per division

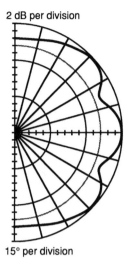

15° per division

Figure 5-20. Lobing response, fourth-order, half an octave above crossover.

2 dB per division

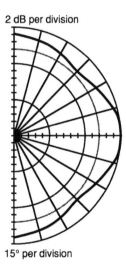

15° per division

Figure 5-21. Lobing response, fourth-order, half an octave below crossover.

The design shown in Figure 5-22*c* combines these virtues. It has lateral symmetry and also provides a graduated set of boundary details for the mid and HF transducers. The designs shown in Figure 5-22*b* and *c* are often made with rounded or tapered edges, thus providing further softening of edges for smoother response.

The worst case here is when a transducer is placed on a baffle so that its

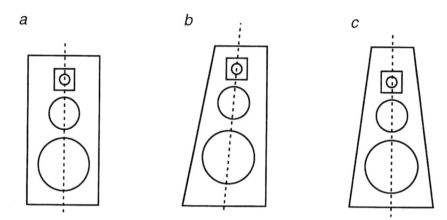

Figure 5-22.   Some typical baffle details. Rectangular enclosure (*a*); trapezoidal enclosure (*b*); truncated pyramidal enclosure (*c*).

distances from the nearest boundaries are equal. The acoustical transmission line path from driver to the multiple edges will produce an impedance discontinuity that will reflect back to the driver, producing an irregularity in response. Distributing these distances, as shown in Figure 5-22*b* and *d*, will minimize the effect.

Figure 5-23 shows some baffle details. The square detail of the typical rectangular enclosure is shown in Figure 5-23*a* in horizontal section view. A strong reflection back to the driver results. Rounding the edge, as shown in Figure 5-23*b* minimizes this. Placing damping material at the edge, as shown in Figure 5-23*c* creates a lossy transmission line boundary and also minimizes reflections. Figure 5-23*d* shows a frontal view of a typical serrated edge treatment often used with damping material.

The shape of the baffle has a profound effect on the response of drivers, as shown in Figure 5-24. In these graphs, a small loudspeaker is mounted as shown in the various objects. The main observations here are the transitions between the 4π and 2π boundary conditions, showing a 6 dB step from low to high frequencies, and the patterns of peaks and dips produced by the various edge details. The data presented here can be scaled as required from the dimensions given.

It is apparent that the spherical enclosure and the truncated pyramidal enclosure offer the smoothest overall response. This accounts for their general popularity in contemporary enclosure design.

Figure 5-25 shows side views of systems consisting of individual enclosures for each transducer (Figure 5-25*a*) and stepped baffle details (Figure 5-25*b*). both techniques are useful in determining the precise location for ideal listening. The stepped baffle approach is usually the more economical of the two.

The array shown in Figure 5-25*c* is often referred to as the d'Appolito array. The design is popular and has an advantage of creating subjective sources for

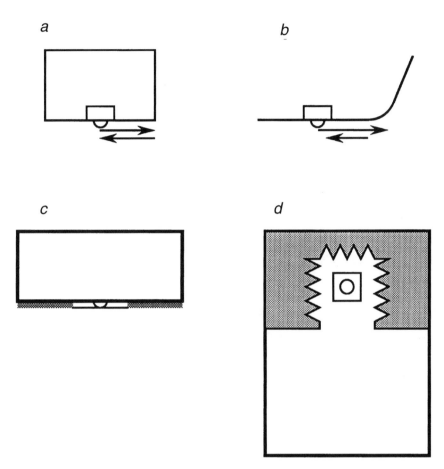

Figure 5-23. Baffle edge details. Rectangular baffle, horizontal section view (*a*); curved baffle, horizontal section detail (*b*); rectangular baffle with damping material, view from above (*c*); front view showing serrated edges on damping material (*d*).

the various bandpass sections that seem to originate from the vertical center point of the array. It is clear that vertical positioning of the listener is critical.

5.7 Time Domain Response of Loudspeakers

Because of the variety of of dividing networks, their specific time domain response, and above all the fore-aft relation between drivers, the overall time domain performance of a loudspeaker system is apt to be anything but uniform over its bandwidth. The steps necessary to provide ideal response may not be absolutely necessary in terms of general audibility, as suggested by the Blauert

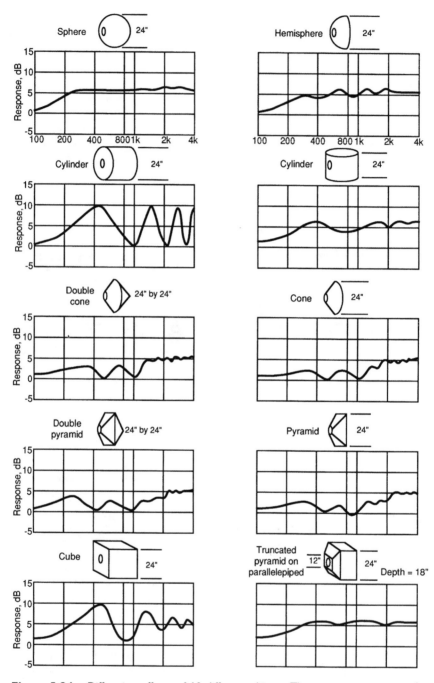

Figure 5-24. Diffraction effects of 10 different objects. The response curves are for small drivers located as indicated on each object.

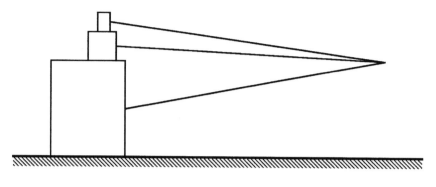

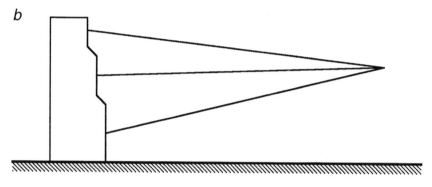

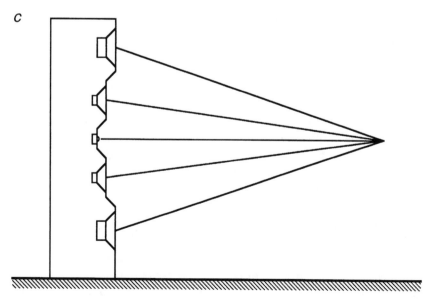

Figure 5-25. Driver displacement fore and aft. Separate enclosures (a); stepped baffle (b); d'Appolito array (c).

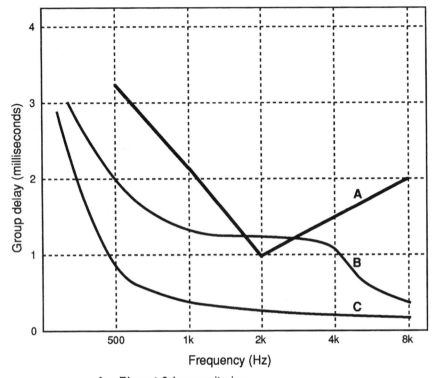

A - Blauert & Laws criterion
B - Typical for 3- way system with midrange horn
C - Typical for 3-way bookshelf system

Figure 5-26. Blauert and Laws criteria for audibility of group delay variations in loud-speaker systems.

and Laws criteria, shown in Figure 5-26. In deriving this data, Blauert and Laws carefully conditioned listening panels to the effects of group delay variations on octave centers over the audio range, and it is generally felt today that the group delay of a loudspeaker system that falls within the envelope will not be audible as such when compared with a system that has no group delay anomalies.

5.8 Loudspeaker Dispersion and Power Response

Most loudspeakers today are forward oriented in response and often exhibit a fairly smooth response over a conical radiation pattern that may be up to 60° wide (this is the included angle between the −6dB zones). Figure 5-27 shows

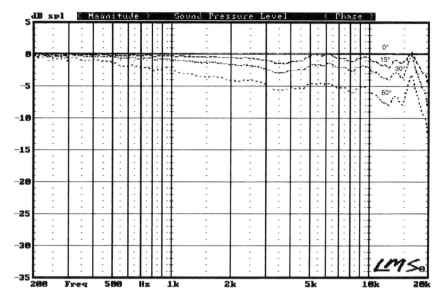

Figure 5-27. Response of a high-frequency horn (normalized to flat on-axis. Averaged ±15°, ±30°, and ±60°, vertical and horizontal, are shown.

how well such response can be maintained with a horn HF element when the horn is designed primarily as a uniform coverage device (see discussion in Chapter 7). As we have seen in earlier chapters, cone drivers exhibit a constant narrowing of response with rising frequency, and any attempt to control the overall coverage angle is apt to involve up to four cone elements to cover the desired frequency range so that the narrowing of pattern can be better controlled. Many engineers feel that it is more important to maintain a gradual narrowing of response with rising frequency rather than a uniform one, particularly if the response is allowed to narrow in the vertical plane while remaining fairly uniform in the horizontal plane.

5.8.1 Power Response Considerations

Strictly speaking, the power response of a loudspeaker is a plot of its power output versus frequency for a given electrical power input. In practice, we do not measure it as such; we infer its relative values from an analysis of on-axis pressure measurements and a knowledge of the DI of the loudspeaker.

The data shown in Figure 5-28a will be useful in determining the DI of cone drivers as a function of frequency. The dashed curve in the figure approximates the directional response of piston radiators mounted on fairly narrow baffles.

The on-axis pressure response and DI of three hypothetical loudspeakers are shown in Figure 5-28*a–c* on one-third octave bands. In each case the pressure

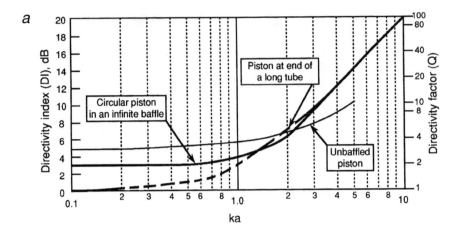

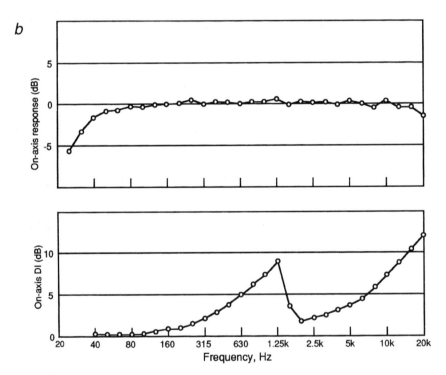

Figure 5-28. On-axis pressure response and DI for three systems. Basic *ka* relationships as a function of driver mounting (*a*); two-way 300-mm LF (*b*); two-way 200-mm LF (*c*); four-way design (*d*).

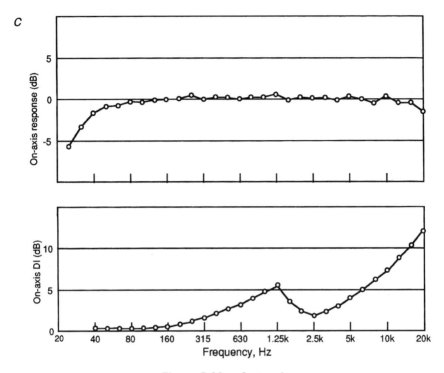

Figure 5-28. *Continued*

response is virtually identical, and what we would like to call attention to is the nature of the DI variations.

The system shown in Figure 5-28*b* is a two-way system with a 300-mm (12-in.) diameter LF unit and a 25-mm (1-in.) diameter HF unit that comes in above about 1500 Hz. This is not a very good design approach, but let us examine the consequences of it nonetheless. Since there is a limit to how low the HF device can be crossed over, we are forced to take the LF unit out to nearly 1600 Hz. At that frequency, the on-axis DI is about 9 or 10 dB.

Once the transition has been made to the HF device, the DI drops down to the range of about 2 dB, from which point it continues to rise with increasing frequency. The rapid variation in the DI curve is the problem. In a very absorptive listening environment, this loudspeaker might sound good on-axis, but in a normal room that had a mixture of both absorptive and reflective surfaces the reflected sound in the room would be strongly influenced by the power response.

The power response is virtually proportional to the *inverse* of the DI curve. This indicates that the power radiated into the room drops off considerably in the 500–1600 Hz range, only to pick up substantially when the HF transition

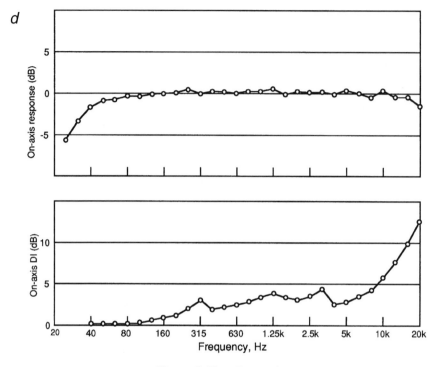

Figure 5-28. *Continued*

has been made. The fact that this variation takes place in the middle of the speech intelligibility range does not help things.

In a two-way design, the cure for this problem is to use as small a LF unit as the system design goals will allow, as we have shown in Figure 5-28c. Here, we are using a 200-mm (8-in.) diameter LF driver, and its DI is only about 7 dB at 1600 Hz. The DI transition with the HF device at crossover is not ideal, but it is far better than in the previous case.

There are many successful "8-in. two-way" systems on the market, often used in conjunction with a subwoofer so as not to incur the obvious output limitations of the 200-mm LF unit.

Obviously, with cone and dome drivers, the only to way to design a system with a fairly uniform DI is to opt for a three- or four-way system. This arrangement is shown in Figure 5-28d. The first crossover takes place at 300 Hz between a 380-mm LF unit and a 200-mm midbass unit. The next frequency division takes place between the 200-mm unit and a 125-mm midrange unit at 1500 Hz. The final transition is to a 25-mm HF unit and takes place at 3500 Hz.

The resulting DI shows a response that climbs gradually from about 300 Hz to the highest frequencies with no major discontinuities. Such a system is expensive and may require that the listener be located along a specific vertical axis. But it will sound to advantage in a variety of listening environments.

In general, if the DI of a loudspeaker system can be maintained within a range of ±3 dB from 250 Hz to about 8 kHz, then it will be relatively free of adverse power response difficulties. Above 8 kHz is it standard practice to let the DI rise as a consequence of normal *ka* relationships.

For the neophyte, as well as experienced loudspeaker engineer, we recommend the excellent design analyses of Dickason (1994). The author clearly presents the complex design processes that are involved, even with the simplest two-way system. The role of the computer is emphasized, both in modeling and in measuring systems. Likewise, Fincham's excellent chapter on multiple-driver loudspeakers in Borwick (1988) is recommended for its thoroughness and technical detail.

Bibliography

Ballou, G., ed., *Handbook for Sound Engineers*, Sams, Indianapolis, IN (1987).

Beranek, L., *Acoustics*, McGraw-Hill, New York (1954).

Borwick, J., *Loudspeaker and Headphone Handbook*, Butterworths, London (1988).

Collums, M., *High Performance Loudspeakers*, Wiley, New York (1991).

Cooke, R., ed., *Loudspeakers, Anthology,* Vols. 1 and 2, Audio Engineering Society, New York (1978 and 1984).

Dickason, V., *Loudspeaker Recipes, Book 1*, Audio Amateur Press, Peterborough, NH (1994).

Jordan, E., *Loudspeakers*, Focal Press, London (1963).

Olson, H., *Acoustical Engineering*, Van Nostrand, New York (1957).

Olson, H., ed., *Loudspeakers, Anthology*, Vol. 3, Audio Engineering Society, New York (1996).

In-Line and Planar Loudspeaker Arrays

6.1 Introduction

While the vast majority of loudspeaker systems are designed around cone or dome transducers, in-line and planar loudspeaker arrays have always occupied a special niche in the high-end consumer market. The electrostatic loudspeaker (ESL) has been highly regarded, and in recent decades has made important strides in sensitivity, power handling, and directional control. In much the same vein, its electromagnetic equivalent, with its printed circuit voice coil, has gained adherents.

Linear arrays, often made up of a large number of small cones or domes, have been on the fringes of loudspeaker system design for some time, and they have now come of age. Electromagnetic ribbon drivers have had a long history but have not enjoyed the popularity of ESLs.

While single cones and domes behave essentially as point sources, exhibiting a 6-dB fall off in level for each doubling of distance from the driver when measured in a free field, line and plane arrays behave very differently, as shown in Figure 6-1a. If we excite a linear array and observe its output level as a function of distance, we will note that the level falls off initially at 3 dB per doubling of distance. This will continue up to some distance A/π, where A is the length of the linear array. Beyond this point the fall off in level will begin to approximate that of a point source, with its characteristic 6-dB per doubling of distance. In a listening environment, a linear array that stretches from floor to ceiling will produce reflected images in both floor and ceiling planes, effectively extending the array vertically. In this environment the 3-dB fall off per doubling of distance may be noticed throughout the listening space.

The attenuation with distance from a plane array is shown in Figure 6-1b. Here, there will be no attenuation until the distance A/π has been reached, at which point a 3-dB per doubling of distance fall off is noticed. This will continue until the distance B/π is reached, beyond which point the familiar 6 dB per

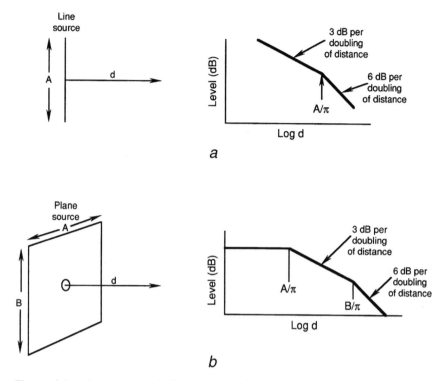

Figure 6-1. Attenuation with distance from a line source (*a*) and a planar source (*b*).

doubling of distance will be noticed. In this representation, *A* is the smaller of the two dimensions of the plane. Again, boundary conditions can alter things considerably. If one wall of a rectangular room is made into a plane loudspeaker array, images in the adjacent planes can create a very large effective plane radiating surface, with little attenuation throughout the listening space. This will create certain problems for stereophonic reproduction, however. In the data shown in Figure 6-1, the transitions between regimes are shown asymptotically; they are actually rather gradual, as observed in the real world.

Much of the appeal of ESLs and their magnetic equivalents comes from their unique radiation properties, as well as their generally low degrees of distortion at moderate drive levels. We begin our discussion with an analysis of the constant-charge ESL design.

6.2 Analysis of the Constant-Charge ESL Loudspeaker

The analysis given here is based on the detailed descriptions given by Baxandall (1988) and Jordan (1963). Readers who wish to learn more about ESLs are encouraged to study these references.

The basic form of the push-pull ESL is as shown in section view in Figure 6-2a. A high dc polarizing voltage is applied to an inner, movable diaphragm, while ground potential is applied to the two outer fixed, perforated electrodes. Under quiescent (no signal) conditions the inner diaphragm is suspended equidistant from the two electrodes. If a signal is impressed on the primary of the transformer so that a positive signal appears on the left electrode, the condition shown at *b* will exist. Since like charges repel, the diaphragm will move to the right, as shown, until balanced by the mechanical restoring force of the diaphragm. Typical values of polarizing electrical fields may be in the range of 20–30 kV/

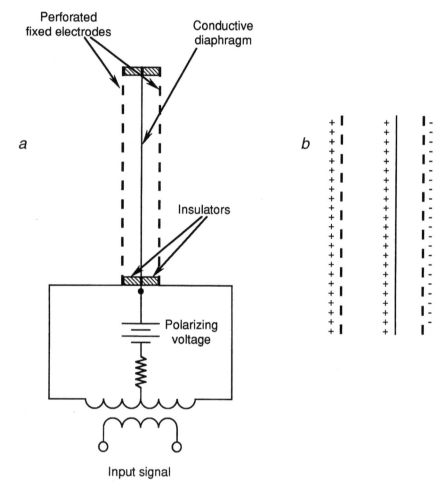

Figure 6-2. The push-pull electrostatic loudspeaker. Section view (*a*); movement of diaphragm for a positive-going signal on the left electrode (*b*).

cm; thus, for a typical model the actual polarizing voltage may be of the order of 2000 or 3000 V.

The governing equation here is:

$$Q = CE = \text{constant} \tag{6.1}$$

where Q is the total charge between the diaphragm and electrodes, C is the total capacitance in the system, and E is the applied voltage. Because of the high value of resistance in series with the diaphragm, the total charge on the plates will not change under normal signal conditions, once the polarizing charging process has reached equilibrium. Therefore, C and E will both vary in inverse relationship in order for their product to be constant.

Under these conditions:

$$Q = AE/4\pi d^2 \tag{6.2}$$

where A is the total area of the diaphragm on both sides (cm^2), E is the polarizing voltage, and d is the distance from the diaphragm to one electrode (in centimeters).

The total force generated then becomes:

$$F = (e_{sig}EA/8\pi d^2)(1.11 \times 10^{-5}) \text{ dyn} \tag{6.3}$$

where E and e_{sig} are in volts, d is in centimeters, and A is in cm^2. Force is thus directly proportional to the applied signal voltage.

A perspective section view of a typical ESL panel is shown in Figure 6-3.

6.2.1 Frequency Response and Directivity

The upper frequency limit of an ESL is generally determined by the high current drawn from the amplifier. Since the loudspeaker load is very largely a capacitive reactance, the current demands and phase angle of the load can be excessive at high frequencies.

Low-frequency displacement limits are generally set by the excursion capability of the diaphragm relative to the fixed electrodes. In terms of acoustical output, the natural LF rolloff of the dipole nature of the system will dominate at the lower frequencies. The larger the radiating area, the lower the useful frequency limit of the system will be.

A single large radiating surface will of course have very irregular HF directional control, as we have seen in polar data for pistons presented in Chapter 1. It is customary to construct the ESL as a two- or three-way system, with the highest frequencies being radiated by a single, relatively narrow vertical strip placed between larger ones for the middle and lower frequencies. In this way, smooth horizontal polar response, with relative freedom from lobing, can be maintained over the normal horizontal listening angle.

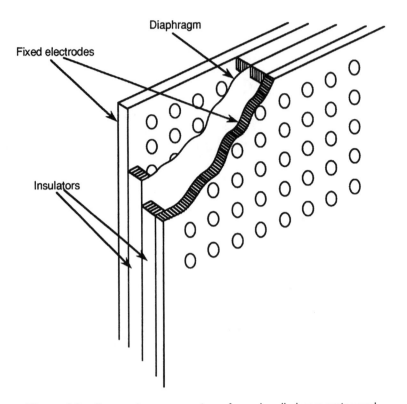

Figure 6-3. Perspective cutaway view of a push-pull electrostatic panel.

A common misconception about ESLs is that their moving mass is extremely small; and therefore that their transient behavior can be exemplary. The fact of the matter is that there is an air layer directly associated with the diaphragm, as discussed in Chapter 1. Jordan (1963) estimates the per-unit area air mass to be about five times greater than the actual per-unit area mass of the diaphragm itself. There may also be a misconception that the diaphragm vibrates essentially with a single degree of freedom; that is, as a unit. The acousto-mechanical impedance of the diaphragm varies over its area, ranging from clamped at the edges to relatively free to move in the middle. Portions of the diaphragm near the edges may be stiffness controlled at high frequencies, while other portions of the diaphragm may not.

There are a number of breakup modes that a rectangular diaphragm can execute, but these are normally damped by viscous air losses through the perforations in the fixed electrodes.

As with the cone driver, the ESL's frequency response depends entirely on its complex radiation impedance and its variation with frequency. Figure 6-4 shows the relative variation in velocity and power radiation per unit area for an

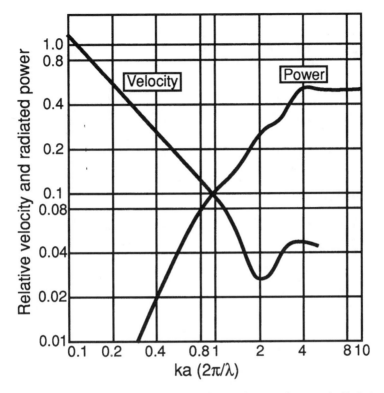

Figure 6-4. Relative values of velocity and radiated power for an unbaffled piston.

ideal unbaffled piston, approximating the acoustical radiation of an ESL. Surface area is all-important with an ESL just as it is for a dynamic driver, and a doubling of radiating area will increase the radiated power by a factor of 4, or 6 dB, for a given frequency and excursion.

6.2.2 Maximum Level Capabilities

In addition to the matter of diaphragm displacement limits, there are electrical limits in level capability of an ESL due to dielectric breakdown between diaphragm and the fixed electrodes. In normal design, it is desirable that the polarizing voltage, plus the maximum intended signal voltage, be slightly less than the breakdown voltage of air itself, so that some kind of safety margin can be held. All insulating materials, plus the coatings on the fixed electrodes should have dielectric constants appropriately high for this application.

The maximum force per unit area that can be generated is:

$$F \text{ (unit area)} = (u^2/16\pi)(1.11 \times 10^{-5}) \text{ dyn/cm}^2 \qquad (6.4)$$

where u is the maximum electrical field strength (V/cm) before the onset of air ionization.

In terms of acoustical pressure output, Walker presents the following equation (Baxandall, 1988):

$$P = I_{sig} \times E/2\pi crd \qquad (6.5)$$

where I_{sig} is the input signal (A), E is the polarizing voltage, c is the velocity of sound (m/sec), r is the measuring distance (m), and d is the diaphragm-electrode spacing (m). P is given in pascals (N/m^2), and the measurement is assumed to be made in the far field.

6.2.3 Details of Construction

In addition to the fairly traditional flat form shown in Figure 6-3, other configurations for ESLs use curved or splayed panels for better control of dispersion, as seen in Figure 6-5. The Quad ESL-63 system uses a set of concentric, sequentially firing, radiating rings in order to achieve effective hemispherical radiation.

6.3 Electromagnetic Planar Loudspeakers

These designs make use of printed voice coil circuits on a sheet of tensioned light plastic, with appropriate layout of ferrite bar magnets arranged in a frame adjacent to the plastic sheet and on one side of it. Openings in the magnet frame provide sufficient egress of sound. Many topologies have been used, and the arrangement shown in Figure 6-6 is typical. The magnets used in these systems are so-called "icebox magnets," similar to those used in just about every home to affix a note or list to the refrigerator door. The field set up by these magnets consists primarily of fringe flux; that is, there is normally no attempt to concentrate the flux density in the area of the conductors through iron magnetic return paths.

Such systems as these can be made in individual sections for HF, MF, and LF coverage in order to achieve good horizontal dispersion. They operate as dipoles at very low frequencies, and in this connection the relatively high resonant Q (low damping) of these systems at LF helps maintain fairly flat response down to the 40 or 50 Hz range. The large expanse of the printed coil may result in a nominal dc resistance in the range of 5 or 6 Ω; thus, there is no need for an additional transformer to match the system with an amplifier. Figure 6-7 shows a photograph of a typical commercial loudspeaker based on these principles.

6.4 The Ribbon Tweeter (High-Frequency Unit)

Details of a typical ribbon HF unit are shown in Figure 6-8. This design was first popularized as the Kelly ribbon tweeter, which was made in England.

Figure 6-5. Photograph of a commercial ESL with curved panels. (Data courtesy Sunrise Audio.)

Normally, such a design is fitted with a small horn to increase its output capability.

Today, much larger ribbon arrays are available, some measuring upward of 2 m. These have the advantage of greater radiating area and hence more output capability. Most of these make use of open magnet structures, as opposed to the arrangement shown in Figure 6-8.

6.5 Discrete Line Arrays

Most line arrays are made up of a number of small cone drivers, making it possible to drive them at different levels and with different frequency response. There is nothing simple about a "simple line array," as we will now observe.

Printed voice coil
on diaphragm

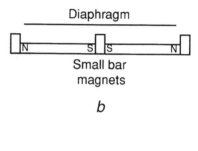

Figure 6-6. Details of an magnetic planar loudspeaker system. View of printed circuit voice coil (*a*); top view of voice coil and diaphragm relative to magnet structure (*b*).

Figure 6-9 shows a perspective view of a vertical column composed of small cone drivers. With the dimensions given, the vertical polar response of this array is shown in Figure 6-10*a–d* for frequencies of 200, 350, 500, and 1000 Hz. For frequencies above 1 kHz, the vertical polar response becomes very sharp on-axis, with the development of many side lobes.

The response for a general line array is given by:

$$R(\phi) = \sin [1/2\ Nkd \sin \phi]/[N \sin (1/2\ kd \sin \phi)] \qquad (6.6)$$

where N is the number of elements in the array, k is equal to $2\pi f/c$, d is the spacing of the elements in the array, and ϕ is the measurement angle in the plane of the array, relative to the normal to the array. The value c is the velocity of

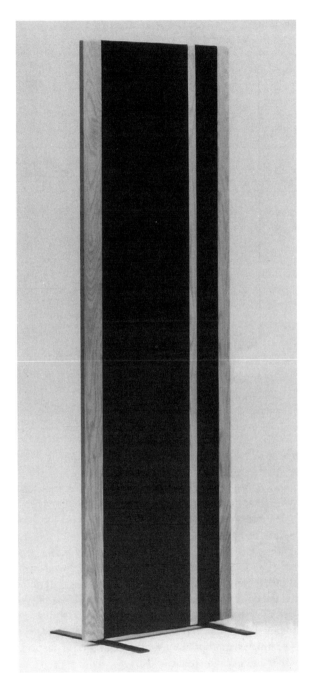

Figure 6-7. Photograph of a magnetic planar loudspeaker system. (Data courtesy Magnepan.)

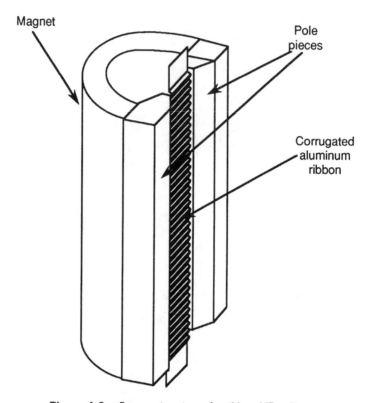

Figure 6-8. Perspective view of a ribbon HF radiator.

sound. All linear measurements are in meters, and the quantities in brackets are to be evaluated in radians.

Figure 6-11 shows the directivity factor for columns composed of more than four elements. Note that for driver separation/wavelength ratios greater than about unity, the on-axis directivity of the array drops as more of the acoustical power is radiated in side lobes.

The chief application of vertical arrays, or "sound columns," is in relatively low-cost speech reinforcement, but for these systems to be useful above the d/λ frequency limit, there must be some kind of frequency shaping, or "tapering," in order to allow the column to become effectively smaller at higher frequencies. Figure 6-12 shows some of the ways this tapering has been accomplished.

The arrangement shown at *a* may be thought of as a column within a column. The inner, HF, section is appropriately shorter than the LF section, and this will provide good coverage at high frequencies.

The arrangement shown in Figure 6-12*b* uses identical drivers, but attains the

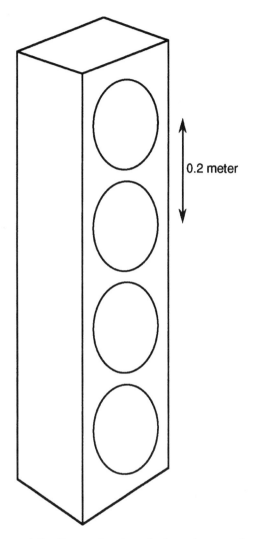

0.2 meter

Figure 6-9. Perspective view of a four-element column.

necessary tapering of frequency response through variable HF damping via wedges of fiberglass (Klepper and Steele, 1963).

The "barber pole" arrangement shown in Figure 6-12c takes advantage of the gradual off-axis fall off of high frequencies in single cone drivers. Observing the column from a fixed horizontal axis, we can see that only the few elements in the middle will be effective at high frequencies, while at middle and lower frequencies progressively more of the drivers will be effective.

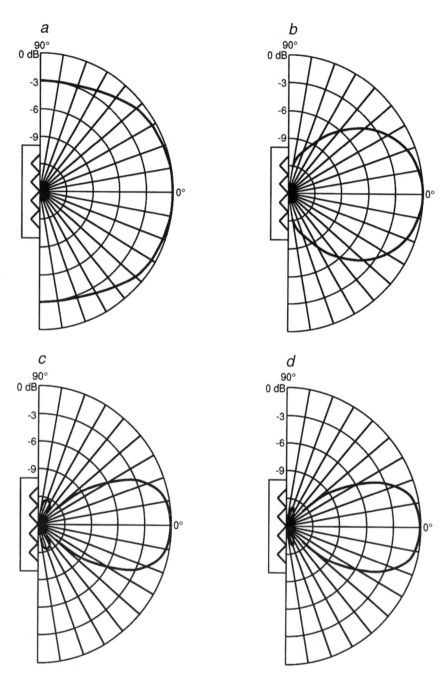

Figure 6-10. Polar response of four-element column; 200 Hz (*a*), 350 Hz (*b*), 500 Hz (*c*), and 1 kHz (*d*).

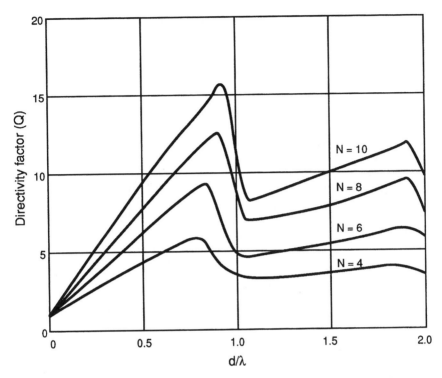

Figure 6-11. Directivity factor plots for simple columns of 4, 6, 8, and 10 elements.

6.5.1 Bessel Arrays

Basing his comments on the work of Franssen at Philips, Kitzen (1983) discusses a family of loudspeaker arrays in which the far-field directional response of an entire line array is very nearly equal to the directional response of a *single driver* in the array. If the Bessel array consists of omnidirectional drivers, the overall directivity in the far field will be omnidirectional. Alternatively, if the array element has some desirable directional characteristic, the array will have basically that same directivity, with the added benefit of increased power handling.

The drive coefficients the elements of the array, and the spacing between drivers, are determined by Bessel coefficients through a design procedure involving complex algebra. We will present here only two such realizations, leaving it to the interested reader to study the Kitzen (1983) reference, which is available as a Philips reprint. In the specific cases shown here, the necessary coefficients can be derived directly from series-parallel hookup of the drivers in the appropriate polarity. Two arrays are shown, one with five drivers (Figure 6-13) and the other with seven drivers (Figure 6-14). The primary application of Bessel arrays is in speech reinforcement systems.

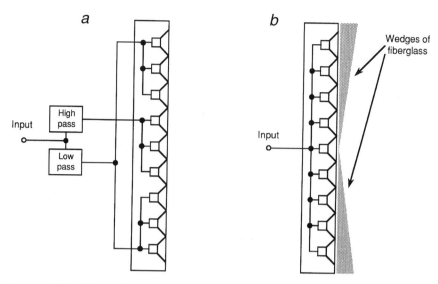

Figure 6-12. Tapered line arrays. Tapering via electrical frequency response (*a*); tapering via acoustical frequency response (*b*); tapering via positional relationships (*c*).

Keele (1990) has analyzed Bessel arrays in considerable detail. In order to show the performance advantages of the five-element Bessel array, Keele presents synthesized polar data on both a five-element standard parallel (non-Bessel) array (Figure 6-15) and the corresponding Bessel array (Figure 6-16). Note the considerably extended frequency range over which the Bessel array provides reasonable omnidirectional response. (In these representations, the length of the array is equal to 1 wavelength at 1 Hz, so that the data can be conveniently re-scaled.)

Keele's analyses further show that the five-element array is the best overall performer in the Bessel family. The price paid for the uniform polar response of the Bessel arrays is the diminished output power capability, as compared with the standard array. In particular, the five-element Bessel is only 16% as efficient as the standard array. However, the bandwidth efficiency product of the five-element Bessel array is excellent, exceeding that of the standard array by slightly more than a factor of 6.

For a five-element Bessel array 1 m in length, rescaling of the operating frequency indicates that excellent response can be maintained well out to 9 kHz.

Keele states that the only significant anomalies in Bessel response are its nonminimum phase performance, both as a function of frequency and measurement angle. These factors would severely limit the degree with which Bessel arrays could be combined with standard arrays.

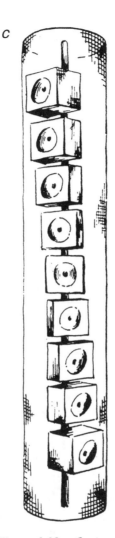

Figure 6-12. *Continued*

6.6 Consumer Systems Using Discrete Element Line Arrays

The Infinity Reference Standard (IRS), shown in Figure 6-17, is a prime example of a multiway line array high-end consumer system incorporating many of the principles discussed in this chapter. High-frequency and MF elements are of the planar printed circuit type arranged vertically. The LF systems are sealed and are response corrected through the use of negative feedback.

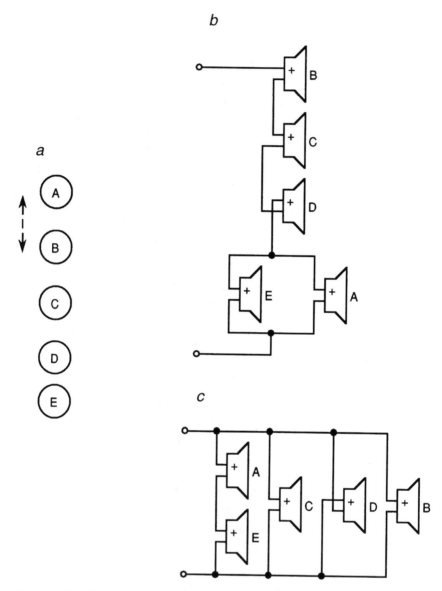

Figure 6-13.   Details of a five-element Bessel array. Array layout (*a*); possible wiring diagrams (*b, c*).

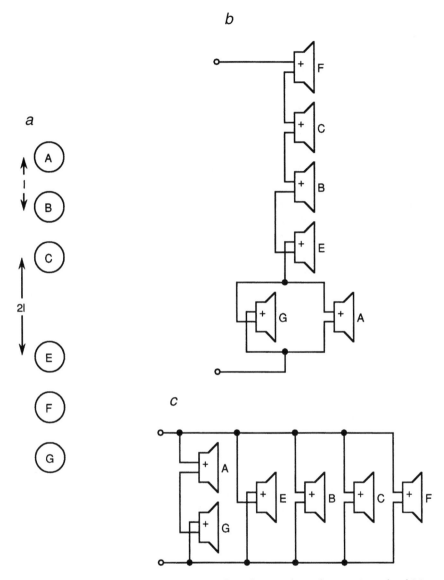

Figure 6-14. Details of a seven-element Bessel array. Array layout; since the driving coefficient for the center loudspeaker is zero, it can be omitted, but normal space for it must be allowed (*a*); possible wiring diagrams (*b, c*).

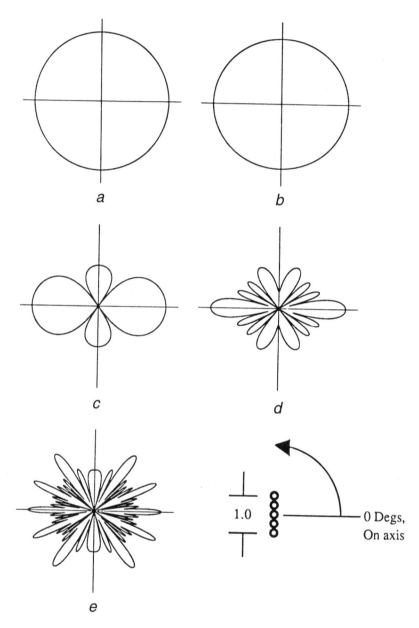

Figure 6-15. Polar response of a five-element standard array (equal level, polarity, and spacing) when array length is 1 wavelength at 1 kHz; 0.1 Hz (*a*), 0.316 Hz (*b*), 1 Hz (*c*), 3.16 Hz (*d*), 10 Hz. Measurements simulated at a distance of 20 times array length; all polars are normalized so that on-axis value is unity. (Data courtesy *J. Audio Engineering Society* and D. B. Keele.)

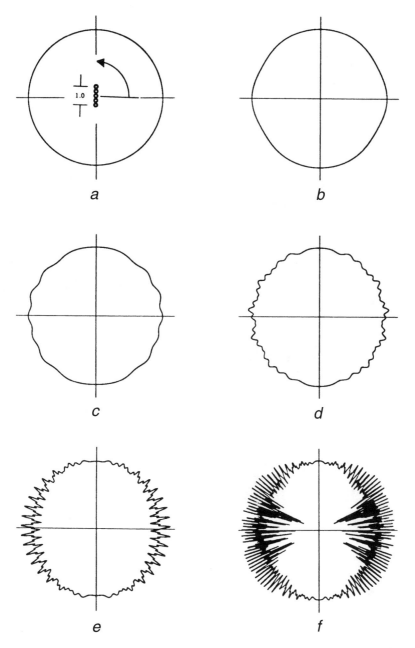

Figure 6-16. Polar response of a five-element Bessel array. Array length is 1 wavelength at 1 Hz; 0.316 Hz (*a*), 1.0 Hz (*b*), 3.16 Hz (*c*), 10 Hz (*d*), 31.6 Hz (*e*), 100 Hz (*f*). Measurements simulated at a distance of 20 times array length; all polars are normalized so that on-axis value is unity. (Data courtesy *J. Audio Engineering Society* and D. B. Keele.)

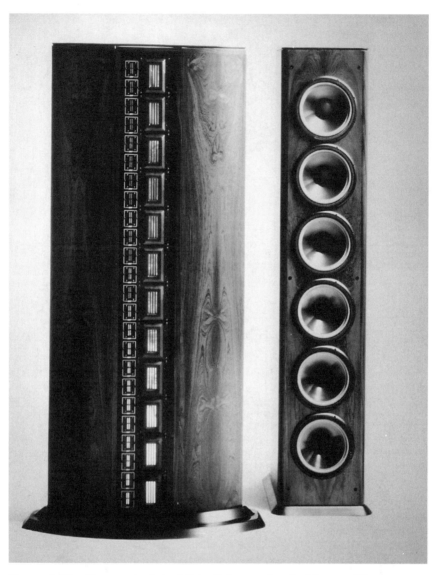

Figure 6-17. Photograph of the Infinity Reference Standard System. (Data courtesy Infinity Systems.)

Bibliography

Baxandall, P., "Electrostatic Loudspeakers," in Borwick, J., ed., *Loudspeaker and Headphone Handbook*, Butterworths, London (1988).

Eargle, J., *Electroacoustical Reference Data*, Van Nostrand Reinhold, New York (1994).

Jordan, E., *Loudspeakers*, Focal Press, London (1963).

Keele, D., "Effective Performance of Bessel Arrays," *J. Audio Engineering Society*, Vol. 38, No. 10 (1990).

Kitzen, J., "Multiple Loudspeaker Arrays Using Bessel Coefficients," *Electronic Components & Applications*, Vol. 5, No. 4 (September 1983).

Klepper, D., and Steele, D., "Constant Directional Characteristics from a Line Source Array," *J. Audio Engineering Society*, Vol. 11, No. 3 (1963).

Kuttruff, H., *Room Acoustics*, Applied Science Publishers, London (1979).

Horn Systems

7.1 Introduction

The history of the horn as an acoustic device dates to antiquity. Early humans used hollowed animal horns for signaling over long distances, and in time the horn became the basis of a number of musical instruments.

Horn loudspeakers were developed fairly early in electroacoustics and were useful primarily because of their relatively high efficiencies and the ease with which their directivity patterns could be controlled. Important work was carried out by Wente and Thuras at Bell Laboratories in the early 1930s and by various manufacturers during the following decades.

For several decades development of horn systems was driven by the requirements of motion picture sound and the need for filling large rooms with high sound pressure levels using power amplifiers of relatively modest output. Further refinements in horn systems during the sixties and seventies were driven by the demands of recording technology and high-level music reinforcement in outdoor venues.

7.2 Horn Flare Profiles

In distinction to a direct radiator, whose moving system is mass controlled and looks into a radiation resistance that rises with frequency, the horn presents a load that is resistive over a large portion of its normal passband. Figure 7-1a shows the section profiles of two horns that have been used in electroacoustics. As shown in Figure 7-1b, the hyperbolic (Hypex) horn (Salmon 1941) exhibits a slight rise in radiation resistance just above the nominal cutoff frequency, f_c, followed by a rapid drop. While it has been used in a number of HF devices, the hyperbolic horn is not generally used today because of its relatively slow flare rate and tendency to develop relatively high amounts of second-harmonic distortion (see Section 7.5).

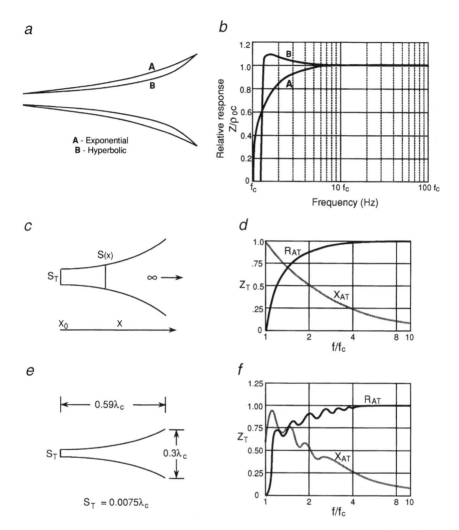

Figure 7-1. Horn profiles and impedances. Section views of exponential and hyperbolic horns (*a*); radiation resistance for exponential and hyperbolic horns (*b*); detail of an infinite exponential horn (*c*); approximate radiation resistance and reactance for an infinite exponential horn (*d*); detail of a finite exponential horn (*e*); radiation resistance and reactance for horn shown in (*e*) (*f*). (Data after Beranek, 1954.)

The profile of the exponential horn is shown in greater detail in Figure 7-1c. The equation for the horn's cross-sectional area, as a function of x, is given by:

$$S(x) = S_T e^{mx} \tag{7.1}$$

where:

$S(x)$ = area at a distance, x, from the throat (m)
S_T = area of the throat (m²)
e = 2.718 (base of the natural logarithm system)
m = flare constant (m⁻¹)
x = distance from the throat along horn axis (m)

The flare constant is:

$$m = 4\pi f_c/c \tag{7.2}$$

where:

f_c = cutoff frequency
c = velocity of sound, m/s

In terms of the horn's cutoff frequency, the approximate complex load at the throat of the horn is given by:

$$Z_{AT} = (\rho_0 c/S_T) \left[\sqrt{1 - (f_c/f)^2} + jf_c/f \right] = R_{AT} + jX_{AT} \tag{7.3}$$

where:

f = driving frequency
$\rho_0 c$ = 406 mechanical ohms at standard temperature and pressure

The values of R_{AT} and X_{AT} for an infinite exponential horn are shown in Figure 7-1d. At frequencies lower than f_c, the impedance of the horn will be largely (but not completely) reactive, and relatively little power will be transmitted. At frequencies much higher than f_c, the reactive term will become very small, and the resistive term will be dominant. For reasons having to do with size, many LF horns are used down to frequencies fairly close to cutoff. HF horns, for reasons having to do with efficiency and proper driver loading, are designed to be used in the range where resistive loading is dominant.

If the circumference of the horn's mouth is greater than approximately three wavelengths of the lowest frequency to be reproduced, then the horn impedance will behave very much like that shown in Figure 7-1d. The data shown at Figure 7-1e and f is for a finite horn of the dimensions indicated. The ripple in the plotted curves is caused by reflections from the mouth of the horn back to the driver due to the acoustical impedance discontinuity the traveling wave sees as it passes the abrupt termination at the mouth. In these figures, λ_c is the wavelength of the cutoff frequency.

The normal bandpass of the horn is that region above which the resistive component of impedance has effectively reached its maximum value, which is equal to:

$$Z_{MT} = \rho_0 c S_T \text{ mechanical ohms} \qquad (7.4)$$

where:

S_T = area of the throat (m^2)

High-frequency horns are often used as high as 8–10f_c and in the range above 4f_c the radiation resistance will be that of a piston, as shown in Figure 1-6. The response will thus show the characteristic ripples in the radiation resistance above $ka = 2$, even in the case of the so-called infinite horn (Keele, 1973).

7.3 The Driving Transducer

The transducer normally used with a horn is called a compression driver, and two views of a modern professional HF driver are shown in Figure 7-2. Drivers such as these provide response from 500 to 800 Hz to about 20 kHz. Figure 7-3 shows a labeled section view of a driver of the type illustrated in Figure 7-2. Figure 7-4a shows details of a relatively low-cost HF driver design, and figures 7-4b and c show details of drivers intended for midrange coverage (from 200 Hz to about 2 kHz).

The HF driver is a precision device with many design tolerances in the range of ±40 μm. The magnetic circuit normally operates with the top plate and pole piece at or near saturation, and gap flux densities in excess of 2.0 T can be attained.

Typical diaphragm diameters range from about 45 to 100 mm. Diaphragm materials have included phenolic-impregnated linen, aluminum, titanium, and beryllium. The metal diaphragms have a thickness of about 40–80 μm. The ideal diaphragm material for extended HF response is one that is rigid, of low mass, and fatigue resistant. The search for better materials continues.

Diaphragm mass is low, and a typical value for a 100-mm diameter diaphragm/ voice coil assembly is about 3.5 g.

The diaphragm is separated from the phasing plug by a space just large enough to ensure that it will not hit the phasing plug on large excursions at lower frequencies. The annular slits in the phasing plug have a collective area that is about one-tenth that of the diaphragm itself, resulting in a pressure-volume velocity transformation ratio of 10 to 1 between the diaphragm and the exit of the phasing plug. It is this mechanical-to-acoustical transformation action that effectively matches the driver to the throat of the horn.

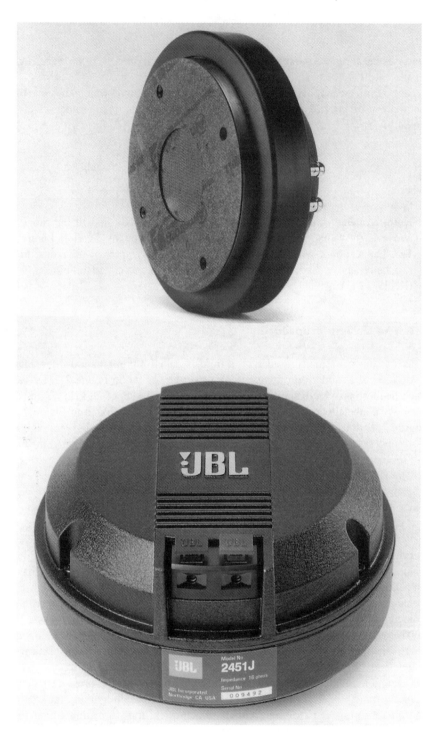

Figure 7-2. Photograph of various HF compression drivers. (Data courtesy JBL, Inc.)

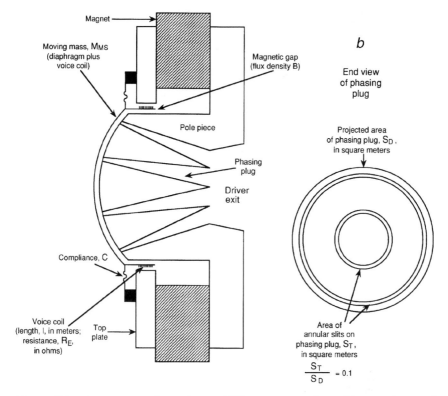

a Section view
of compression driver

Magnet

Moving mass, M$_{MS}$
(diaphragm plus
voice coil)

Magnetic gap
(flux density B)

b

End view
of phasing
plug

Pole piece

Projected area
of phasing plug, S$_D$,
in square meters

Phasing
plug

Driver
exit

Compliance, C

Voice coil
(length, l, in meters;
resistance, R$_E$,
in ohms)

Top
plate

Area of
annular slits on
phasing plug, S$_T$,
in square meters

$$\frac{S_T}{S_D} = 0.1$$

Figure 7-3. Section view of a professional HF compression driver (*a*); normal view of phasing plug, diaphragm side (*b*). (Data courtesy JBL, Inc.)

7.3.1 Analysis

An analogous circuit for the driver, loaded by a horn, is shown in Figure 7-5*a*. On the mechanical and acoustical side, the circuit is of the mobility type, where M_{MS} is the mass of the moving system (diaphragm and voice coil). C_{MS} is the mechanical compliance and C_{MB} is the compliance of the air space behind the diaphragm; r_{MS} and r_{MB} are their associated values of mechanical responsiveness. C_{MI} is the compliance of the small (but by no means insignificant) airspace between the diaphragm and the phasing plug.

In the bandpass of the driver we can simplify this equivalent circuit, reflecting it back to the electrical side as shown in Figure 7-5*b*. Here, R_E represents the electrical resistance in the amplifier-voice coil circuit and R_{ET} represents the

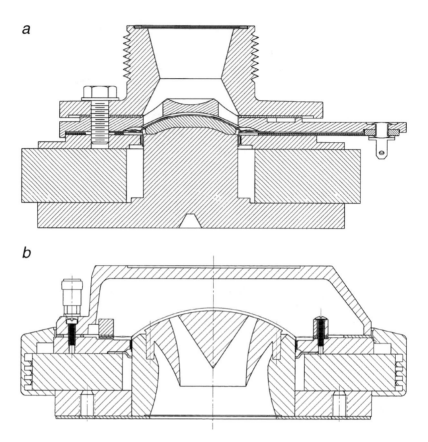

Figure 7-4. Section view of a low-cost HF compression driver (*a*); section view of the JBL Model 2490 midrange compression driver (*b*); section view of the Community M4 midrange compression driver (*c*). [Data courtesy JBL, Inc. (*a,b*); Community (*c*).]

effective radiation resistance in ohms reflected through the mechanical and acoustical systems:

$$R_{ET} = S_T(Bl)^2/\rho_0 c S_D^2 \qquad (7.5)$$

If a driver has been designed for flattest response, R_E and R_{ET} will both be just about equal, and the efficiency will be:

$$\text{Efficiency} \ (\%) = \frac{2\ R_E R_{ET}}{(R_E + R_{ET})^2} \times 100 \qquad (7.6)$$

This value will be 50% when R_E and R_{ET} are equal.

At high frequencies, the equivalent electrical circuit is as shown in Figure

c

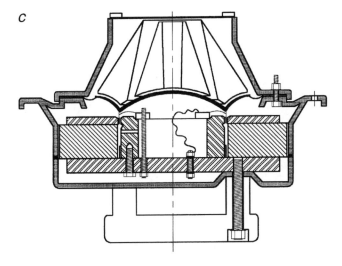

Figure 7-4. *Continued*

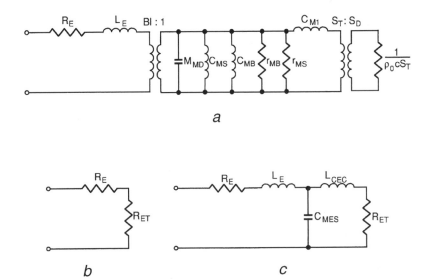

Figure 7-5. Equivalent circuit of a compression driver (*a*); simplified circuit for midband operation (*b*); simplified circuit for HF operation (*c*).

7-5c. The additional reactive elements affect the frequency response by progressively rolling off HF response. L_E represents the inductance of the voice coil, and its effect can always be seen in the impedance curve of the driver as a rise in impedance at higher frequencies. Some driver manufacturers deposit a thin copper or silver ring on the polepiece, which minimizes the rise in impedance due to the coil's inductance by acting as a transformer with a shorted secondary turn.

C_{MES} is a shunt capacitance which causes the so-called mass breakpoint in the driver's frequency response. This is a 6 dB/octave rolloff in HF response commencing at f_{HM}:

$$f_{HM} = \frac{(Bl)^2}{\pi R_E M_{MS}} \tag{7.7}$$

The mass breakpoint is significant in all HF compression drivers and normally falls in the range between 3 and 4 kHz.

Finally, L_{CEC} is a series inductance corresponding to the front air chamber between the diaphragm and phasing plug. In some drivers the effect of the front air chamber may be noticed at about 8 kHz. For drivers optimized for high frequencies its effect may be negligible.

At low frequencies, the driver's response is limited by the primary diaphragm resonance, and in most HF drivers this usually takes place in the range of 500 Hz. Although a given horn may provide extended resistive loading below the driver's resonance, operating the driver below resonance will call for careful monitoring of signal input. Most manufacturers will state a driver power derating value for such operation.

7.3.2 The Plane Wave Tube (PWT)

In carrying out compression driver development, engineers normally measure the driver's response on a plane wave tube (PWT), not a horn. Figure 7-6a shows details of the PWT. The tube is of the same diameter as the exit of the driver, and a probe microphone is placed fairly close to the mounting flange for the

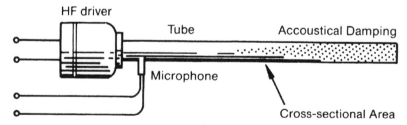

Figure 7-6.   The plane wave tube (PWT). Section view of PWT (*a*); plots of amplitude response and impedance for JBL 2445J driver on PWT (*b*) (Data for (*a*) courtesy JBL, Inc.)

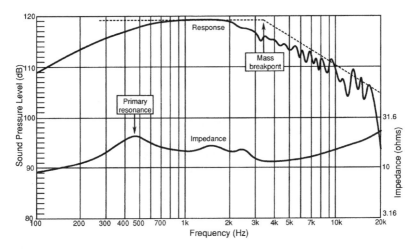

Figure 7-6. *Continued*

driver. There is a progressive loss of acoustical power as sound progresses down the tube, which is caused by a tapered wedge of fiberglass or other suitable damping material. The tube may be 2 or 3 m in length, and by the time sound has propagated over that distance, it has become attenuated to such a point that there is little acoustical power to reflect back to the microphone. Acoustically, the tube presents a resistive load to the driver such as would be presented by an infinite horn with a cutoff frequency of zero Hz.

The sound pressure level in the tube is uniform in the portion ahead of the damping material and is equal to:

$$L_p = 94 + 20 \log \sqrt{W_A(\rho_0 c)/S_T}, \qquad (7.8)$$

where W_A is the acoustical power in watts delivered by the driver and S_T is the cross-sectional area of the tube in square meters.

Most manufacturers normalize their PWT data to a standard tube with a diameter of 25.4 mm (1 in.). Let us now put 1 W of acoustical power into such a tube and calculate the value of L_p:

$$L_p = 94 + 20 \log\sqrt{(1)406/0.0005}$$
$$L_p = 94 + 20 \log (901) = 94 + 59 = 153 \text{ dB}$$

As a rule, a reference power of 1 mW is used, producing a level of 123 dB in the tube.

Figure 7-6*b* shows the typical 1 mW PWT response of a JBL 2445 compression driver. Note that the maximum level in the bandpass of the device is 119 dB L_p. This value is 4 dB lower than the 1 mW reference level of 123 dB L_p, indicating that the driver's efficiency is $100 \times 10^{-4/10}$%, or 40%.

Furthermore, we can see the effect of the mass breakpoint in the driver's response in the 3.5 kHz range. We can compare this with the value given by equation (7.4):

$$Bl = 18 \text{ T}$$
$$R_E = 8.5 \ \Omega$$
$$M = 0.00346 \text{ kg}$$
$$f_{HM} = (18)^2/\pi(8.5)(.00346) = 3506 \text{ Hz}$$

On the graph, a breakpoint at 3500 Hz has been superimposed by dotted lines over the response curve. Note that it effectively matches the transition point in the curve.

The modulus of impedance of the driver has been plotted below the response curve, and the driver resonance at about 500 Hz is obvious.

7.3.3 Secondary Resonances in the Compression Driver

Our analysis thus far has assumed that the driver's diaphragm moved as a unit. At high frequencies, other forms of motion become obvious and may have a profound effect on response. There is, in most drivers, a secondary resonance that occurs in the surround, or suspension, of the driver. When this resonance takes place, motion from the voice coil causes considerable motion in the surround itself and relatively little in the diaphragm. There is usually a peak in the overall response, followed by a pronounced HF rolloff.

Figure 7-7 shows superimposed response curves of three drivers mounted on

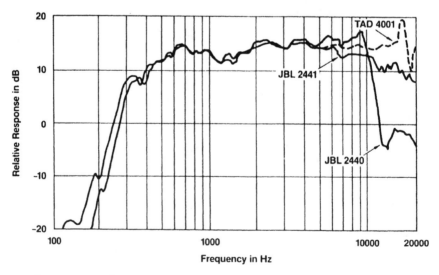

Figure 7-7. Effect of horn loading of a driver, compared with PWT loading.

a JBL 2350 radial horn. The curves are unequalized, and two of them (JBL 2440 and TAD 4001) show the effect of secondary resonances. The elevated response of the 2440 driver at 9 kHz and at 17 kHz in the 4001 driver are due to secondary resonances at those frequencies. The beryllium diaphragm in the 4001 driver is considerably stiffer than the aluminum diaphragm in the 2440, and this accounts for the shift of nearly an octave in the secondary resonance. By comparison, the 2441 diaphragm design has suppressed those resonances, allowing a smoother overall curve, but one that has more apparent rolloff than the other two in the range between 5 and 10 kHz. These response curves are typical of the many design factors and trade-offs that the transducer engineer must deal with.

7.4 Ring Radiators

Ring radiators are very-high-frequency devices operating normally in the range from about 4 to 16 kHz. They embody the elements of a compression driver and horn in a single unit. A section view of the JBL 076J ring radiator is shown in Figure 7-8. The diaphragm is clamped at the outer edge as well as in the middle, effectively forming an annular, or ring-shaped, radiating surface. The outer portion of the diaphragm has greater mass than the inner portion, and this gives rise to a double tuned moving system operating over a range of about two octaves. The initial horn flare is annular in shape, eventually making a transition into a horn with elliptical cross section. Other models of the ring radiator have the same basic driving mechanism but with different horn configurations.

Over their operating range, ring radiators can exhibit efficiencies on the order of 6.3% and handle power inputs of up to 40 W, yielding a power output, per device, of about 2.5 acoustic watts. They are often used in multiple arrays for greater output.

7.5 Families of Horns

In this section we will discuss the various families of horns that have been developed over the last half-century of loudspeaker evolution. The exponential horn basically has poor pattern control, narrowing progressively with rising frequency. Each in its own way, the families of horns we will discuss here attempted to solve problems in pattern control, while maintaining some degree of exponential flare for good loading. We will discuss them in the approximate order in which they were developed.

7.5.1 Multicellular Horns

The multicellular horn was developed during the 1930s and found extensive application in motion picture theaters, where it was necessary to aim sound specifically at seating areas in large houses. It also became a staple in sound

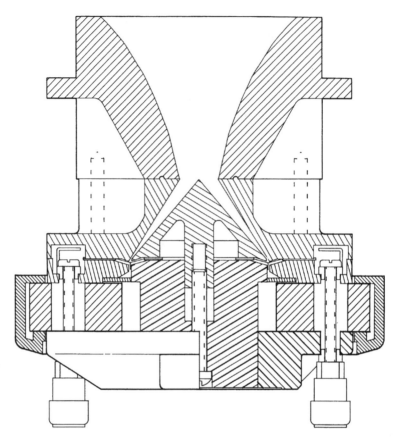

Figure 7-8. The ring radiator. Section view of JBL model 2402 UHF driver. (Data courtesy JBL, Inc.)

reinforcement applications during the 1950s, but was later generally replaced by the radial horn.

Figure 7-9a shows a photograph of a group of multicellular horns. Individual cells were nominally 25 degrees wide and were of exponential flare. Models with 800 and 500 Hz lower frequency limits were the most common. The theoretical pattern control of the multicellular horn is shown in Figure 7-9b. Note that pattern control above about 1 kHz is determined by cell multiples; however, in the range between 500 Hz and 1 kHz, there is substantial narrowing of the pattern as the distance across the mouth of the assembled cells increases. At lower frequencies the cell array loses pattern control altogether.

Selected polar measurements are shown in Figure 7-9c–e. At 1 kHz, as seen in Figure 7-9c, the patterns are well behaved and follow the data shown in Figure 7-9b. At 2 kHz, shown in Figure 7-9d, the polar response is still good, but a

a

Figure 7-9. The multicellular horn. Group photograph of multicellular horns (*a*); theoretical beamwidth data for multicellular horns (*b*); vertical and horizontal polar response of horn at 1 kHz (*c*), 2 kHz (*d*); 10 kHz (*e*). (Photograph courtesy of Altec Lansing.)

pattern of "fingering" can be seen toward the edges of the patterns. At 10 kHz, shown in Figure 7-9*e*, there is considerable fingering in the main coverage zones of the horn, even to the extent of 10-dB variations in the horizontal plane. Note that fingering is at a maximum along the septum dividing adjacent cells.

7.5.2 Radial Horns

Figure 7-10*a* shows a photograph of a family of radial horns. Top and side views of a typical design are shown in Figure 7-10*b*. The radial horn is so named because its straight sides outline radii of a circular arc formed by the horn's mouth. They are also called sectoral horns. As seen in side view, the horn's flare is exponential; as seen in top view it is conical. In the vertical plane the exponential flare determines the pattern control, and in Figure 7-9*c* we note that the pattern narrows progressively with rising frequency. However, in the horizontal plane, the pattern remains fairly constant, corresponding to the angle set by the straight sides. At frequencies above 8 or 10 kHz there may be horizontal pattern variations due to interference problems in the early flare development at the throat. Directional properties of the JBL 2345 radial horn are shown in Figure 7-10*c*.

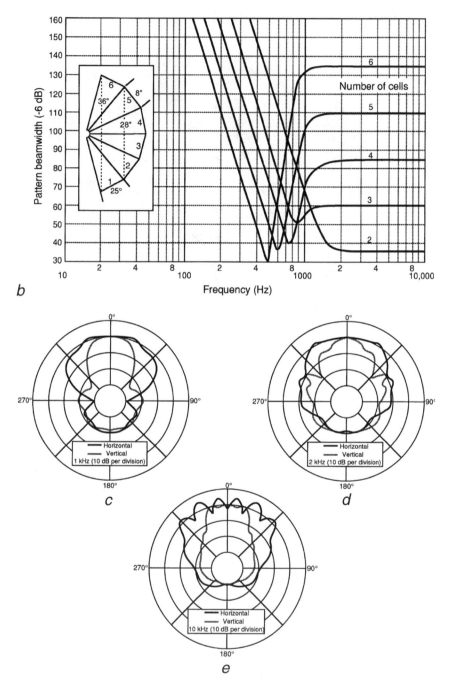

Figure 7-9. *Continued*

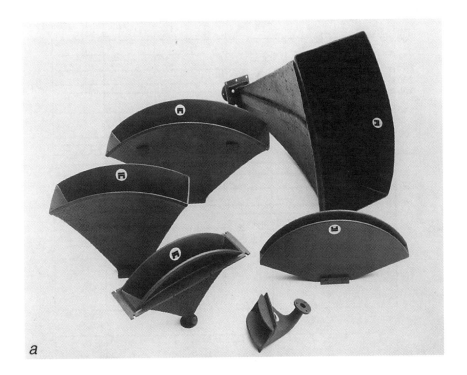

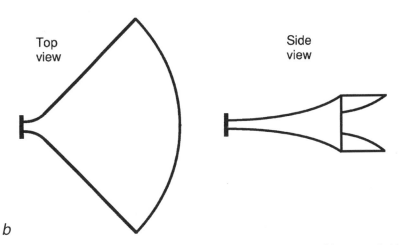

Top view · Side view

Figure 7-10. The radial horn. Photograph of various radial horns (*a*); top and side drawings of a radial horn (*b*); directivity data for JBL model 2345 horn (*c*). (Data courtesy JBL, Inc.)

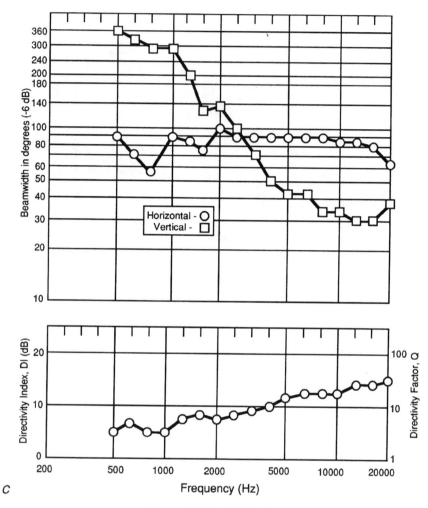

C

Figure 7-10. *Continued*

The horizontal pattern loses control below that frequency whose wavelength is approximately equal to the mouth width. Often, there is a narrowing of the horizontal pattern in this range, just above the frequency where control is lost.

For applications in which vertical pattern control may not be critical, the radial horn is an excellent choice, and many modern studio monitoring systems are designed with them. The rising directivity index acts as an acoustical equalizer, boosting the on-axis response and counteracting, to some extent, the driver's rolloff above the mass breakpoint.

7.5.3 Acoustic Lenses

Acoustical lenses came into prominence during the 1950s, when they were first used in motion picture systems. Like the multicellular horn, they were developed essentially to compensate for the HF beaming of exponential horns.

Figure 7-11*a* shows a photo of a family of horn-lens combinations. There are two basic lens types, slant plate and perforated plate. Details of the slant plate lens are shown in Figure 7-11*b*, and the perforated plate is shown in Figure 7-11*c*. The directional characteristics of a small format slant plate lens is shown in Figure 7-11*d*.

The lens works by providing a shorter sound path through the middle than at the sides. Thus, the wavefront exits the device with greater curvature and hence greater dispersion. The slant plate lens provides wide pattern control only in one plane, while the perforated plate lens spreads high frequencies in a conical pattern. Directional properties of the JBL 2307/2308 slant plate lens are shown in Figure 7-11*d*.

The acoustical lens is little used today, but the designs are held in high esteem by many users. They were the hallmark of many of the JBL studio monitors during the 1960s and 1970s.

7.5.4 Diffraction Horns

A diffraction horn has a mouth that is quite narrow in one plane and fairly wide in the other, as shown in Figure 7-12*a*. Pattern control in the plane perpendicular to the narrow opening will be quite wide at middle and lower frequencies, since it is largely diffraction controlled. In the other plane pattern control is dictated by the mouth width. In the case of the horn illustrated here, the mouth defines an angle of 120°, and it is necessary to incorporate a set of tapered guides inside the horn to maintain wide-angle response at high frequencies. The directional characteristics are shown in Figure 7-12*b*.

Diffraction horns have had considerable application in monitoring systems over the years, and variations of this design are very much in use today.

7.5.5 Uniform Coverage Horns

In studying the directional data on the horns so far discussed, it is apparent that none of the horns exhibits uniform control in both horizontal and vertical planes. As a general rule, control in the vertical plane has been sacrificed for desired pattern control in the horizontal plane.

During the 1970s and early 1980s, considerable work was done by Henricksen and Ureda (1978) and Keele in designing horns that were able to maintain uniform pattern control in both planes. Such devices are universally used today for general

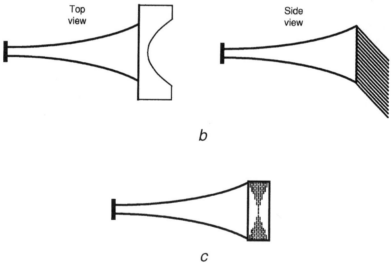

Figure 7-11. The acoustic lens. Photograph of various horn-lens combinations (*a*); side and top view drawings of a slant plate horn-lens (*b*); section view of a perforated plate horn-lens (*c*); directivity data for JBL 2307/2308 horn-lens combination (*d*). (Data courtesy JBL, Inc.)

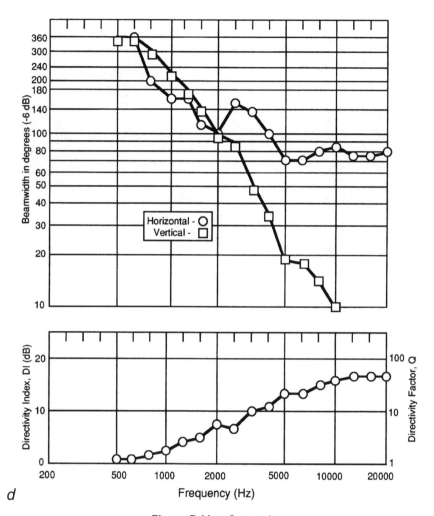

d

Figure 7-11. *Continued*

sound reinforcement and motion picture work. They are also known as *constant coverage* horns.

Figure 7-13*a* shows a photo of a group of uniform coverage horns. Detailed top and side views of a 90°-by-40° uniform coverage horn are shown in Figure 7-13*b* and *c*. In the horizontal plane (90° coverage), a curved diffraction slot feeds a large flared bell whose sides have been tapered to maintain the desired coverage angle to 10 kHz and beyond. The initial section of the horn which feeds the diffraction slot is exponential, and the flare rate is determined by the requirement of attaining the desired vertical coverage angle.

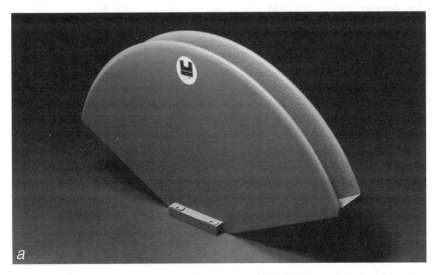

a

Figure 7-12. The diffraction horn. Photograph of JBL 2397 horn (*a*); directivity data for 2397 horn (*b*). (Data courtesy JBL, Inc.)

In the vertical plane (40° coverage), the horn appears to be nearly straight-sided, except for a slight flare at the mouth. Directional characteristics are shown at *d*. Note that the pattern control in both vertical and horizontal planes is smooth over the frequency decade from 1 to 10 kHz. In the horizontal plane the pattern is maintained down to 500 Hz.

Uniform coverage horns have changed the way sound system engineers equalize their systems. Figure 7-14*a* shows a family of on- and off-axis response curves for the JBL 2350 radial horn. The on-axis curve (0°) reflects the response of the driver as modified by the on-axis DI of the horn. As can be seen from Figure 7-10*c*, the DI for a typical radial horn rises with frequency. This rise in DI adds directly to the driver's response, as measured on the 0° axis. As we see here, that curve is quite flat from 800 Hz to about 10 kHz, and no further horn-driver equalization will be required.

However, with uniform coverage horns, the relatively constant DI, as seen in Figure 7-13*d*, requires that the mass breakpoint fall off of the driver's response be electrically boosted for flat on-axis response. This process is shown in Figure 7-14*b*. The result is a near-ideal set of on- and off-axis curves, ensuring a very uniform frequency response over the horn's horizontal 90° included angle.

7.5.6 Influence of Mouth Size on Low-Frequency Pattern Control

If a horn is to maintain desired pattern control down to some given frequency, the horn's cutoff frequency will have to be well below the desired coverage

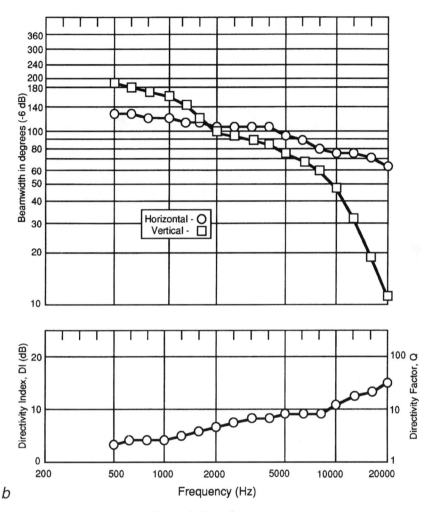

Figure 7-12. *Continued*

frequency, and the horn's mouth will have to be of a certain minimum dimension in order to maintain the pattern. A relatively small mouth dimension will be able to maintain, say, a 90° −6 dB beamwidth at a given frequency; however, in order to maintain a beamwidth of, say, 40° at the same frequency, a much larger mouth dimension will be required.

The data presented in Figure 7-15*a* shows the approximate behavior of a 90°-by-40° horn. We can identify clearly the frequencies at which the nominal −6 dB horizontal and vertical beamwidths have doubled. These are labeled, respectively, α and β in the figure. It has been observed that horns with rectangular apertures exhibit a consistent loss of pattern control as given in Table 7.1.

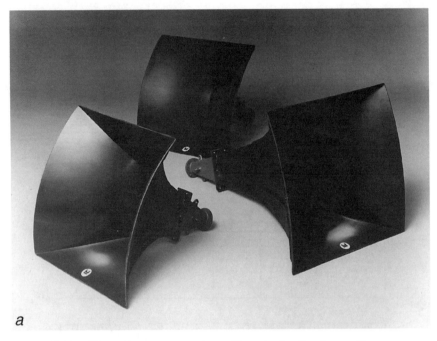

a

Figure 7-13. The uniform coverage horn. Photograph of various uniform-coverage horns (*a*); top view of JBL model 2360 horn (*b*); side view of 2360 horn (*c*); directivity data for the JBL 2360 horn (*d*). (Data courtesy JBL, Inc.)

Table 7-1. Pattern Control as a Function of Coverage Angle and Mouth Dimension

Nominal Coverage Angle	Nominal Angle Doubles At
90°	λ = 3 times mouth height
60°	λ = 2 times mouth height
40°	λ = 4/3 times mouth height
20°	λ = 2/3 times mouth height

This data has been plotted as shown in Figure 7-15*b*. For example, let us find the minimum mouth width for a horn which will maintain a 90° angle down to 500 Hz. The pattern doubles approximately at 250 Hz. Locate 250 Hz on the abscissa and move upward to the 90° line. Read the value of 500 mm on the ordinate. For a nominal 40° coverage angle at 500 Hz, the mouth width will have to be approximately 1 m.

7.5.7 Horn Impedance Measurements

Figure 7-16 shows superimposed plots of the impedance modulus for the JBL 2445 driver mounted on a PWT and mounted on the JBL 2380 horn. Such

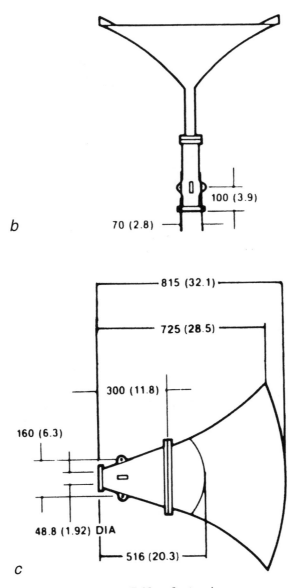

Figure 7-13. *Continued*

variation in impedance is typical of horns, and it bears more than a slight resemblance to the data shown in Figure 7-1*d*. In general, a 3-to-1 variation in impedance is acceptable over the bandpass of the driver-horn combination. Variations in excess of this usually indicate abrupt transitions in the horn flare itself and may be reflected as variations in the system's amplitude response.

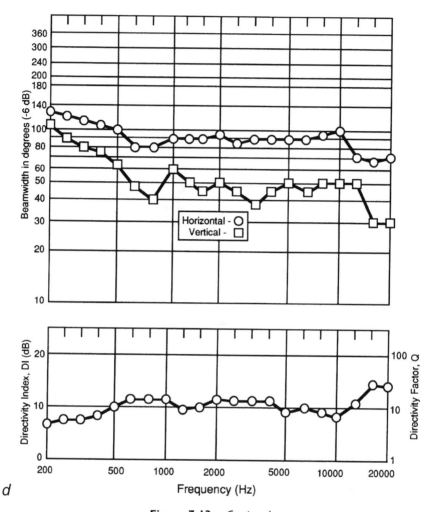

Figure 7-13. *Continued*

7.6 Distortion in Horn Systems

The predominant form of distortion in a well-designed horn system is second-harmonic distortion, which comes as a result of high pressures in the horn's throat. The effect is thermodynamic in origin and can only be reduced by increasing the flare rate so that pressures downstream in the horn are more quickly relieved. Figure 7-17*a* shows the nonlinearity in the volume-pressure relationship for air

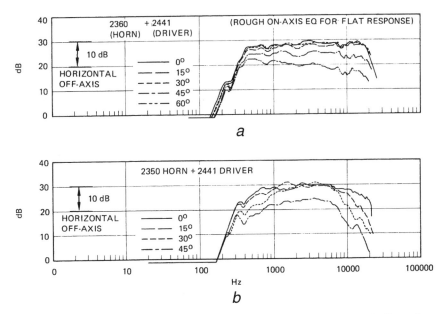

Figure 7-14. Uniform coverage: new hardware versus the old. On- and off-axis frequency response for JBL 2350 radial horn (*a*); on- and off-axis frequency response for JBL 2360 Bi-Radial uniform coverage horn, equalized for flat power response (*b*). (Data courtesy D. B. Keele and JBL, Inc.)

that causes the distortion. The following equation (Thuras, 1935) gives the value of second-harmonic distortion:

$$\text{Percent second harmonic} = 1.73\ (f/f_c)\sqrt{I_T} \times 10^{-2} \tag{7.9}$$

where f is the driving frequency, f_c is the horn cutoff frequency, and I_T is the intensity at the throat in watts per square meter. Figure 7-17*b* presents graphical solutions to this equation.

Figure 7-17*c* shows second-harmonic distortion for two driver-horn combinations. For these tests, the fundamental output was maintained flat from 1 to 10 kHz, and the measured second-harmonic distortion was raised 20 dB for ease in reading.

We can compare these measurements with calculations using Equation (7.9), as follows.

The JBL 2360/2446 horn/driver combination has a cutoff frequency of 70 Hz and a midband sensitivity of 113 dB L_p, 1 W measured at 1 m. In these tests, the level, referred to 1 m, was set 6 dB lower for 107 dB at 1 m for an electrical power's input of 0.25 W. As read from the driver's PWT data in Figure 7-6, the

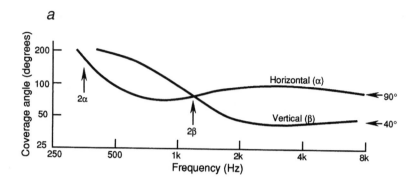

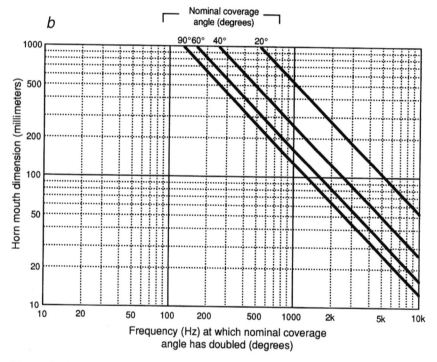

Figure 7-15. Horn mouth size versus −6 dB beamwidth control. Approximate beamwidth behavior of a 90°-by-40° horn (*a*); horn mouth dimension required for desired beamwidth control (*b*).

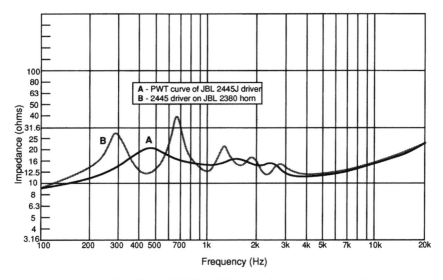

Figure 7-16. Effect of PWT and horn loading on driver impedance.

driver has an efficiency of 40%, so we now know that the acoustical power generated by the driver is $(.25)(.4) = 0.1$ W.

The exit diameter of the driver is 50 mm, which has an area of 1.96×10^{-3} m². The intensity is then $0.1/(1.96 \times 10^{-3}) = 51$ W/m². Thus, at 2 kHz:

$$\text{Percent second} = 1.73 \ (2000/70) \ \sqrt{51} \times .01 = 3.52\%$$

Checking the data in the graph, we see that the measured distortion reads just below 3%, indicating excellent agreement between calculation and measurement (a difference of about 1.3 dB).

The JBL 2352 horn has a much higher cutoff frequency in both driver and horn flare development, resulting in distortion 6–8 dB lower than the 2360 horn.

7.7 Horn Driver Protection

In many applications, HF horn systems carry a large portion of the acoustical load, with very high crest factors when HF power response equalization is added to the signal input. Also, in other applications, there may be LF surges when amplifiers are turned on. Either effect here can cause driver damage and care must be taken that the devices are not unduly stressed.

Low-frequency protection can be provided by placing a high-quality capacitor in series with the driver using values taken from Table 7.2.

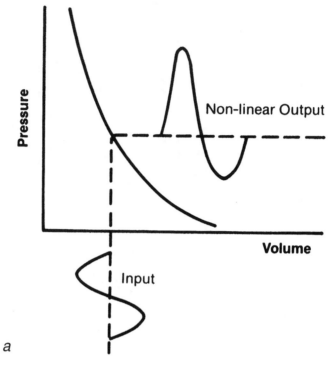

a

Figure 7-17. Second-harmonic distortion in horn systems. Basic thermodynamic action (*a*); distortion as a function of acoustic intensity at the throat versus ratio of driving frequency to cutoff frequency (*b*); plots of distortion for two horn-driver combinations with different cutoff frequencies (*c*).

Table 7-2. Blocking Capacitor Values for Horn Driver Protection Capacitor Value

	For Protection Below These Frequencies (Hz)		
	16 Ω	8 Ω	4 Ω
72 μF	275	550	1,000
52 μF	400	750	1,500
20 μF	1,000	2,000	4,000
16.5 μF	1,200	2,500	5,000
13.5 μF	1,500	3,000	6,000
12 μF	1,700	3,500	7,000
8 μF	2,500	5,000	10,000
7 μF	3,000	6,000	12,000
6 μF	3,500	7,000	14,000
4 μF	5,000	10,000	—
3 μF	7,000	14,000	—

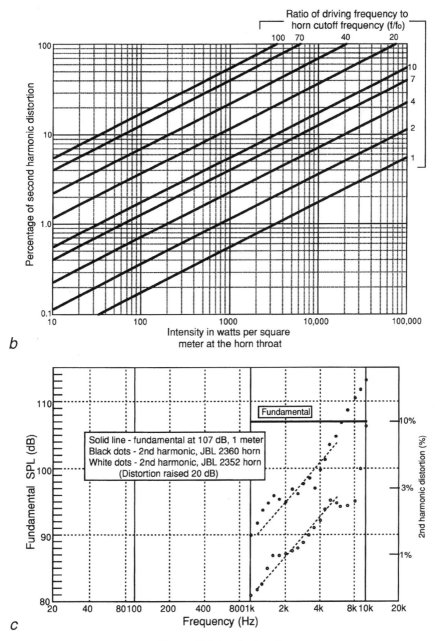

Figure 7-17. *Continued*

The values are selected to present a reactance at one-half of the crossover frequency equal to the nominal impedance of the driver.

High-frequency limiting can be used for driver protection, and if properly applied its audible effects will be very slight, if noticeable at all. The methods shown in Figure 7-18 have been used for passive driver protection at high frequencies. The method shown in Figure 7-18*a* and *b* uses back-to-back Zener diodes chosen to clamp the signal to the driver at the desired voltage maximum values. Consult the driver manufacturer for details before doing this. The effects may be audible for prolonged application.

The bridge circuit shown in Figure 7-18*c* was developed by ElectroVoice and

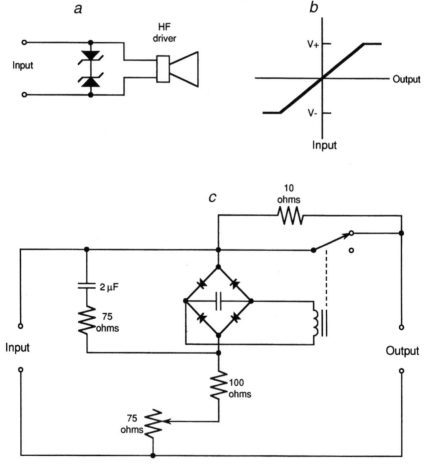

Figure 7-18. High frequency driver protection. Zener diodes (*a*) can be used to clamp the signal to a safe voltage (*b*); a relay can be used to reduce the level of HF signals fed to the driver (*c*).

provides a rectified voltage for operating a relay in series with the HF driver. When the relay is actuated a 10-Ω resistor is placed in series with the driver, reducing the signal to a safer level.

7.8 Low Frequency Horns

The biggest problem today with LF horns is their size and expense. The earliest examples were for use in motion picture theaters, as typified by the W-horn developed by Hilliard and Clark (1938) shown in Figure 7-19.

Perhaps the most sophisticated LF horn design is that of Paul Klipsch (1941), the so-called Klipschorn, shown in Figure 7-20. This is a 40-Hz horn that attains an extended mouth dimension through corner placement and ingenious folding of the early flare development.

Like many designs, the Klipschorn takes advantage of reactance annulling at very low frequencies. The air chamber behind the LF driver provides mechanical stiffness that effectively cancels the mass reactance presented by the complex impedance of the horn. This enables the resistive impedance component to be significant down to the range of horn cutoff. The design provides uniform response from 40 to 400 Hz, with significant output down to 32 Hz.

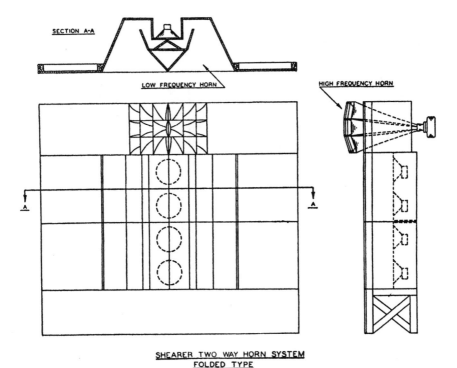

SECTION A-A

LOW FREQUENCY HORN

HIGH FREQUENCY HORN

SHEARER TWO WAY HORN SYSTEM
FOLDED TYPE

Figure 7-19. Views of a W-horn used during the 1930s in motion picture applications.

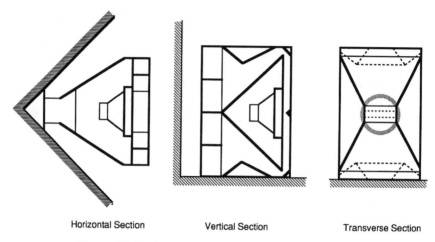

Horizontal Section   Vertical Section   Transverse Section

Figure 7-20. Section views of the Klipschorn LF section.

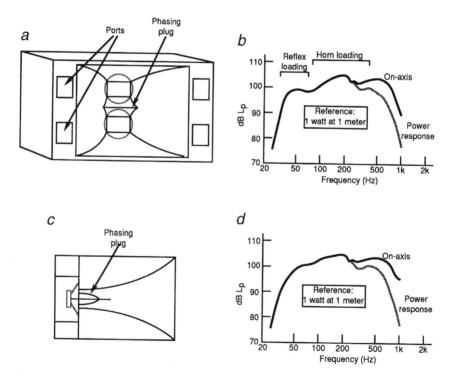

Figure 7-21. Bass horns. JBL 4550 ported horn enclosure (*a*); typical response of the 4550 system (*b*); section view of a straight bass horn (*c*); typical response (*d*).

7.8.1 Hybrid Designs and Straight Low-Frequency Horns

Many of the LF horns used in high-level music reinforcement today are, strictly speaking, not horn loaded at the lowest frequencies because they have a cone-throat area ratio of unity. Rather, the horn mouth and sides are used for some degree of directional control in the very important 150–300 Hz range. Typical here would be the ported horns that were developed during the 1940s for motion picture application, similar to that shown in Figure 7-21*a* and *b*. Note that the response curve shows two distinct effects, reflex loading of the drivers at the lowest frequencies and horn loading of the drivers in the mid-bass range.

Straight LF horns have generally been used in music reinforcement. Many of these are of moderate size and are used primarily for directional control in the midbass down to the 50–60 Hz range. Details are shown in Figure 7-21*c* and *d*.

7.8.2 Driver Parameters for LF Horn Application

Keele (1977) developed the data shown in Figure 7-22 for determining the various response breakpoints with a given set of Thiele-Small LF parameters. As seen in the figure, there will be a region of flat power response bounded by f_{LC} at low frequencies and by f_{HM} at higher frequencies. Beyond f_{HM} there will be two

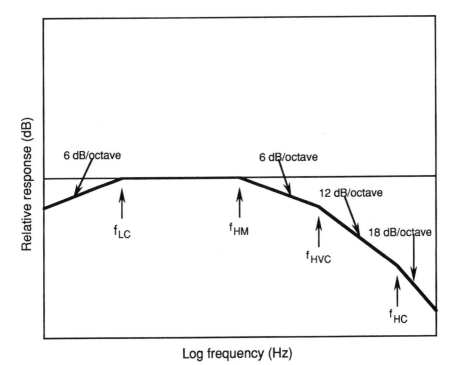

Figure 7-22. Use of Thiele-Small driver parameters in optimizing LF horn systems.

additional breakpoints in the response envelope due to voice coil inductance and the effect of the air chamber directly in front of the transducer's diaphragm. Each of these breakpoints will add a 6 dB per octave rolloff to the overall response curve. The various breakpoints are defined as follows:

$$f_{LC} = (Q_{ts})f_s/2 \qquad\qquad (7.10)$$

$$f_{HM} = 2(f_s)/Q_{ts} \qquad\qquad (7.11)$$

$$f_{HVC} = R_e/\pi L_e \qquad\qquad (7.12)$$

$$f_{HC} = (2Q_{ts})f_s(V_{as}/V_{fc}) \qquad\qquad (7.13)$$

where:

Q_{ts} = total Q of transducer
f_s = free-air resonance of the transducer
R_e = voice coil dc resistance (Ω)

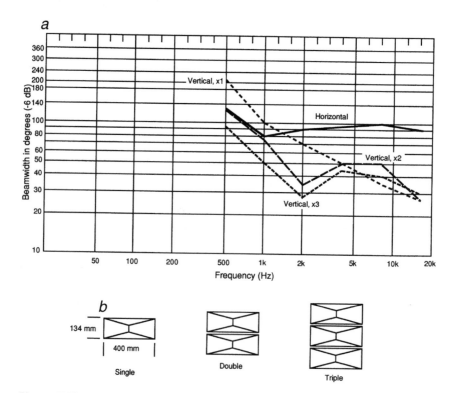

Figure 7-23. High-frequency horn stacking to provide higher directivity in the vertical plane. Beamwidth data (*a*); stacking configurations (*b*).

L_e = voice coil inductance of the transducer (H)
V_{as} = volume of air that provides a restoring force equal to that of the driver's
    mechanical compliance (L)
V_{fc} = volume of front air chamber (L)

7.9 Horn Arrays

Horns are very useful in many professional sound applications because precise directional properties can be designed into them. Furthermore, horn arrays can be assembled both to narrow coverage angles as well as to widen them. Some of the options here are shown in Figure 7-23, where vertical stacking of horns can be used to increase low and mid-frequency directivity in the vertical plane. If this technique is taken beyond three horns, lobing in the vertical plane may defeat the advantages that have been gained (JBL, 1984).

The beamwidth data is shown in Figure 7-23*a* and the three stacking versions are shown in Figure 7-23*b*. For speech reinforcement applications, the reduced

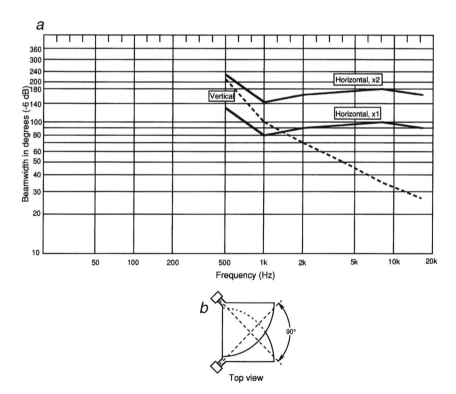

Figure 7-24. Stacking and splaying of horns to increase horizontal coverage. Beamwidth data (*a*); stack-splay configuration (*b*).

directivity in the 2 kHz per octave range will improve speech intelligibility because of the increase in direct-to-reverberant ratio.

The data shown here is for the JBL small-format 2345 radial horn. Similar results may be scaled for different horn mouth dimensions.

Horns may be stacked and splayed along their −6-dB coverage zones to increase the horizontal coverage, as shown in Figure 7-24. Beamwidth data is shown at *a*, and the stack-splay configuration is shown in Figure 7-24*b*. Both horns individually have 90° horizontal coverage, so they have been splayed by 90° for a total horizontal coverage angle of 180°.

Bibliography

Beranek, L., *Acoustics*, Wiley, New York (1954).

Frayne, J., and Locanthi, B., "Theater Loudspeaker System Incorporating an Acoustic Lens Radiator," *J. Society of Motion Picture and Television Engineers*, Vol. 63 (September 1954).

Henricksen, C., and Ureda, M., "The Manta-Ray Horns," *J. Audio Engineering Society*, Vol. 26, (1978), No. 9.

Hilliard, J., and Clark, L., "Headphones and Loudspeakers," in *Motion Picture Sound Engineering*, Van Nostrand, New York (1938).

JBL, Inc., Technical Note 1(7), Northridge, CA (1984).

JBL, Inc., Technical Notes 1(8), Northridge, CA (1985).

Johansen, T., "On the Directivity of Horn Loudspeakers," *J. Audio Engineering Society*, Vol. 42, No. 12 (1994).

Keele, D., "What's So Sacred about Exponential Horns?" presented at the 51st Convention of the Audio Engineering Society, *J. Audio Engineering Society* (abstracts) Vol. 23, p. 492 (1975); Convention Preprint 1038.

Keele, D., "Low Frequency Horn Design Using Thiele-Small Driver Parameters," Presented at Audio Engineering Society Convention, Los Angeles, May 1977; Preprint 1250.

Keele, D., "Optimum Horn Mouth Sizes," presented at the 46th Audio Engineering Society Convention, New York, September 1973; Convention Preprint 933.

Klipsch P., "A Low-frequency Horn of Small Dimensions," *J. Acoustical Society of America*, Vol. 18 (1941), pp.137–144.

Lansing, J., and Hilliard, J., "An Improved Loudspeaker System for Theaters," *J. Society of Motion Picture Engineers*, Vol. 45 (1945), pp. 339–349.

Leach, M., "A Two-port Analogous Circuit and SPICE Model for Salmon's Family of Acoustic Horns," *J. Acoustical Society of America*, Vol. 99 (1966), No. 3, pp. 1459–1464.

Locanthi, B., "Application of Electric Circuit Analogs to Loudspeaker Design Problems," *J. Audio Engineering Society*, Vol. 19 (1971), p.778–785.

Olson, H., *Acoustical Engineering*, Van Nostrand, New York (1957).

Plach, D., "Design Factors in Horn-Type Speakers," *J. Audio Engineering Society*, Vol. 1, No. 4 (1953).

Salmon, V., "Hypex Horns," *Electronics*, Vol. 14 (1941), p. 39.

Wente, E., and Thuras, A., "Auditory Perspective—Loudspeakers and Microphones," *Electrical Engineering*, January 1934.

Thuras, A., "Extraneous Frequencies Generated in Air Carrying Intense Sound Waves," *J. Acoustical Society of America*, Vol. 6 (1935), pp. 173–180.

Electronic Interface

8.1 Introduction

In this chapter we will deal with the interface between amplifier and loudspeaker. Subjects will include the detailed nature of the loudspeaker load, amplifier characteristics, line losses, parallel and series operation of both amplifiers and loudspeakers, multi-amping, and loudspeaker protection methods.

8.2 The Power Amplifier

The great majority of amplifiers used today are solid state, or transistor, designs. The art has progressed steadily over decades, and a very high order of performance is now available. Nevertheless many audiophiles still prefer vacuum tube designs, either older or newer models.

Figure 8-1 shows a simplified view of one channel of a modern solid-state stereo amplifier. In normal operation it is essential that both output voltage and current be constrained to certain limits. As opposed to tubes, transistors will fail catastrophically if certain limits are exceeded, even if only for very brief times.

An oscilloscope can be used to monitor both the the input and output of an amplifier, as shown in Figure 8-2. Here, the input signal is fed to the horizontal (x-axis) deflection plates in the oscilloscope and the output signal is fed to the vertical (y-axis) deflection plates. In this example, the output stages have been driven into clipping, and the ± output voltage limits are clearly shown.

We can also use the oscilloscope to observe output voltage and output current relationships, as shown in Figure 8-3. If the load on the amplifier is resistive, current and voltage will be in phase with each other, and a diagonal line will be traced on the face of the scope, as shown in Figure 8-3a. If the load is both resistive and reactive, as many loudspeakers are, the current-voltage relationship will be elliptical, as shown in Figure 8-3b. The actual phase angle may be computed from the ratio of height of the ellipse and the y-axis intercept as:

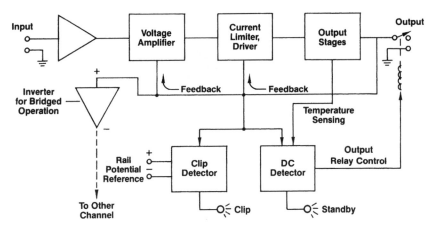

Figure 8-1. Simplified view of one-half of a stereo solid-state power amplifier. (Data courtesy JBL, Inc.)

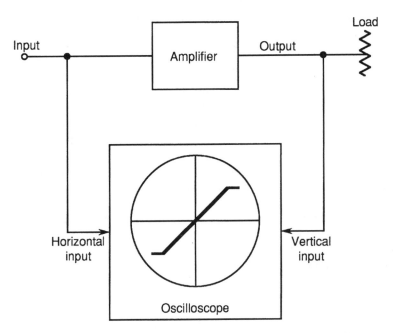

Figure 8-2. Monitoring input and output of an amplifier with an oscilloscope.

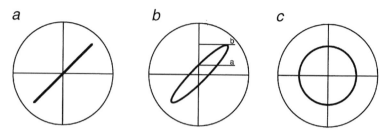

Figure 8-3. Scope figures. Current and voltage with a resistive load (*a*); a load with both resistive and reactive components (*b*); a load that is purely reactive (*c*).

$$\text{Phase angle} = \cos^{-1}(a/b) \qquad (8.1)$$

If the load is purely reactive, such as a capacitor, the scope trace will be circular, indicating that voltage and current are 90° with respect to each other, or in quadrature. Under this condition, no net power is transferred to the load. Power fed to the load on one half-cycle is returned to the amplifier on the other half-cycle.

The signal returned to the amplifier from the reactive element on the negative half-cycle is referred to as back electromotive force (EMF), or back voltage. In the case of high-efficiency loudspeaker systems, the back EMF may cause trouble, in that it may force the amplifier to pass current when the output voltage is zero. This is a problem that is usually discussed under the subject of safe operating conditions for the amplifier.

8.2.1 Amplifier Current and Voltage Limits

An amplifier is designed with fixed output current and voltage limits, and there are additional limits based on the instantaneous relation between the two. Early solid-state amplifiers often had little or no protection against short-circuited outputs, and transistor failure was common. Later amplifiers have solved these problems, often, some would say, at the expense of sonic integrity. The better engineered an amplifier is, the more tolerant it will be of adverse loading, going in and out of its protection modes as needed with little degradation of sound.

From a loudspeaker design viewpoint, every effort should be made to keep the load from becoming too reactive or too low in resistance. As a target, the phase angle of the load should never exceed ±60°, and the lowest resistive load value presented by the loudspeaker over its operating passband should not be less than about 80% of the nominal load value.

8.2.2 The Basic Design Impedance

We can define the impedance at which the amplifier can deliver its maximum amount of power:

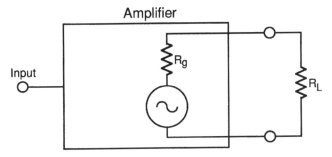

Figure 8-4. Measurement of amplifier-load damping factor. R_g = Amplifier generator (or source) resistance; R_L = Load resistance.

$$Z = E_{max}/I_{max} \qquad (8.2)$$

Most professional amplifiers today will have a design impedance in the range of 4 Ω or less, since this is normally about the lowest value of impedance at which an engineer would choose to operate a system.

8.2.3 Damping Factor

For most practical purposes we can assume that the source impedance of a modern power amplifier is virtually zero. That is, the output section behaves as a constant voltage source, regardless of variations in the load itself. Strictly speaking, there is a finite source impedance looking back into the amplifier, and the damping factor is defined as:

$$\text{Damping factor} = R_L/R_g \qquad (8.3)$$

where R_L is the nominal value of the load and R_g is the source resistance of the amplifier's output section. Details are shown in Figure 8-4. Values of damping factor in the range of 200–250 are common. Damping factor is measured by loading down the amplifier's output with increasingly low resistance until the output voltage has dropped to one-half of its open-circuit value. When this condition is reached, the value of the load equals that of the internal source impedance of the amplifier.

8.3 Line Losses

Much more significant than amplifier source resistance is the effect of line resistance between the amplifier and load. Tables 8.1 and 8.2 show values for different gauges of copper wire, both solid and stranded. In most professional sound installations today, it is standard practice for power amplifiers to be located at or very near their respective loudspeakers, thus ensuring minimum line losses.

Table 8-1.   Resistance for Various Gauges of Solid Copper Wire

Conductor cross-sectional area (mm²)	AWG[a] number	Resistance per 300 m (1000 ft) pair (Ω)
5.2	10	2.00
3.3	12	3.15
2.1	14	5.00
1.3	16	8.00
0.87	18	12.5
0.52	20	20.0

[a]American Wire Gauge.

Line losses may be calculated by the method shown in Figure 8-5. The equation to be solved here is:

$$\text{Loss (dB) at the load} = 20 \log [R_L/(R_L + 2R_1)] \tag{8.4}$$

Note that the loss at the load in decibels as given here results from two effects; one is the actual loss in the line and the other is the loss due to the target impedance mismatch at the load itself.

A great deal of money can be spent on system wiring, perhaps more than really need be. One beneficial fallout of expensive cable is the generally high quality of the connectors themselves. In this regard, it is a good thing to periodically disconnect a system and reconnect it, making sure that all connections are positive and clean. Any areas showing signs of corrosion should be thoroughly cleaned.

The reader should also bear in mind that the loudspeaker itself has a dc resistance that may be in the range of 5 or 6 Ω. Common sense indicates that an additional series resistance of perhaps one-hundredth that value should have little if any audible effect on the loudspeaker's performance. However, if line losses are excessive, say, in the range of 1 or 2 Ω, the predominant audible effect

Table 8-2.   Resistance for Various Gauges of Stranded (Tin-Coated) Copper Wire

Diameter of stranded conductor (mm)	AWG[a] number	Resistance per 300 m (1000 ft) pair (Ω)
3	10	2.54
2.6	12	4.16
1.9	14	6.0
1.5	16	9.5
1.3	18	12.1
1.0	20	18.1

[a]American Wire Gauge.

American Wire Gauge (AWG #)	Resistance per single run 300 meters (1000 feet) of copper wire (ohms)
5	.3
6	.4
7	.5
8	.6
9	.8
10	1.0
11	1.2
12	1.6
13	2.0
14	2.5
15	3.2
16	4.0
17	5.0
18	6.3
19	8.0
20	10

Note: Paralleling two identical gauges reduces effective gauge by 3.

EXAMPLE: Find the power loss at an 8-ohm load due to a fifty meter run of AWG #14 wire

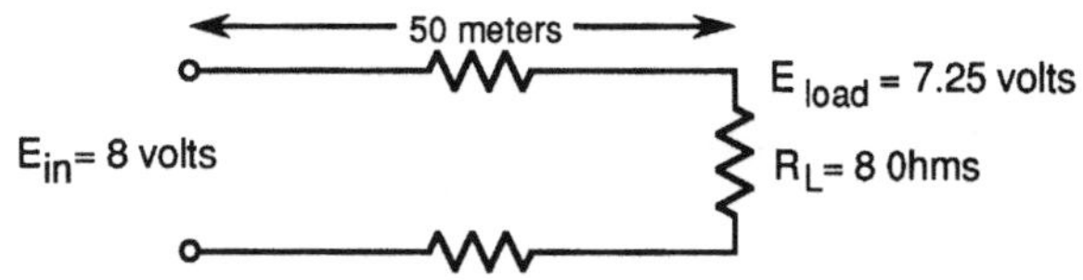

$$R = \frac{50}{300} \times 2.5 = 0.416 \text{ ohms}$$

$$E_{load} = \frac{8}{8 + (2 \times .416)} \times 8 = 7.25 \text{ volts}$$

$$\text{Power in load} = \frac{(7.25)^2}{8} = 6.56 \text{ watts}$$

$$\text{dB loss} = 10 \log (6.56/8) = .86 \text{ dB}$$

Figure 8-5. Calculation of loudspeaker line losses.

may be a change in the frequency balance of the loudspeaker, due to the interaction of line resistance with the varying impedance of the loudspeaker with respect to frequency.

8.3.1 Series and Parallel Hookup of Loudspeakers

Where many loudspeakers are to be operated by a single amplifier, as in paging or music distribution systems, it is customary to use a series-parallel arrangement of loads so that the net impedance is in the proper operating range of the amplifier. While such practice may be acceptable for noncritical applications, it is not recommended for demanding applications. Each loudspeaker should ideally look back directly into the power amplifier for proper electromagnetic damping.

For professional distribution systems, the 70-V line is often used. In this arrangement, the full output of the power amplifier is available at 70 V root-mean-square (rms). Individual distribution transformers are then used with each driver to tap a given amount of power from the line. The user does not have to keep track of load impedances; rather, the total amount of power drawn from the line is summed and of course must not exceed that available from the amplifier. For further discussion of this subject the reader is referred to Chapter 11.

8.4 Matching Loudspeakers and Amplifiers

An age-old question in audio engineering is what size power amplifier to use with a given loudspeaker. As we have seen earlier, the loudspeaker system carries a thermal power rating that is based on cumulative heating effects. The system may also carry a displacement power rating (more aptly, a derating) for very-low-frequency (VLF) application.

Taking these ratings at their face value, we can come up with a rational approach to power amplifier selection based on the system applications. There are three general cases:

1. *High-level music reinforcement applications.* The recommendation here is that an amplifier be used that has a continuous output power rating exactly equal to the thermal rating of the loudspeaker. This will ensure that a normal, highly compressed musical program will drive the loudspeakers safely at all times.

2. *Speech reinforcement applications.* The recommendation here is that an amplifier be chosen that has a continuous power output rating equal to twice the thermal rating of the loudspeaker system. This will give an additional 3 dB of program headroom, which the loudspeakers can take in stride on a momentary basis.

3. *Critical music monitoring applications.* The recommendation here is that an amplifier be chosen that has a continuous power output rating equal to four times the thermal rating of the loudspeaker system. This will give an additional

6 dB of program headroom, to be judiciously implemented by the supervising engineer only when peak program demands require it.

It is strongly recommended that amplifiers for professional use be chosen that have high current capability, inasmuch as certain dynamic program conditions can give rise to peak current values that are far greater than those that would be predicted from an examination of the loudspeaker's complex impedance curves based on the conventional continuous swept sine wave measurement signal (Otala and Huttunen, 1987).

8.4.1 Pertinent Amplifier Specifications

Outside of the nominal power rating of an amplifier and its damping factor, primary specifications are: voltage sensitivity, input impedance, distortion at rated output (with all recommended loading options), and output noise. Secondary specifications may deal with dynamic headroom capability and matters of power bandwidth.

The voltage sensitivity of an amplifier is the input rms voltage that will produce rated output power into a stated load value. A typical value here would be 1 V rms for rated output into 8 Ω.

Input impedances are normally resistive and values in the range of 20,000 Ω are common.

Distortion at rated output power is normally stated for the 20 Hz to 20 kHz range, as well as 1 kHz, for each recommended load impedance. Wide-range values of 0.15–0.2% are common, while values at 1 kHz may be as low as 0.015–0.02%. These measurements are normally made into a purely resistive load.

Output noise is normally a wide-band measurement, and values of 100 dB below rated output are common.

8.5 Amplifier Bridging

Many stereo amplifiers today can routinely be set up for bridged operation. In this mode, the two amplifiers are effectively operated in series, and the output voltage capability is thus doubled. The basic hookup is shown in Figure 8-6. The two output grounds are connected together, and the amplifiers are driven in opposite polarity. (Note the detail in Figure 8-1 showing an inverting amplifier stage for bridged operation.)

A very important point in bridged operation is that the load impedance must be *twice* that for the amplifiers when used in their normal mode. For instance, assume that a stereo amplifier is rated for continuous operation of 600 W per channel into 4 Ω. In the bridged mode the new rating would be 1200 W into 8 Ω.

Bridging is a convenient way of matching amplifiers with a 4-Ω design impedance to drive loads of 8 Ω or higher. The problem is that two stereo amplifiers will be needed for normal stereo operation.

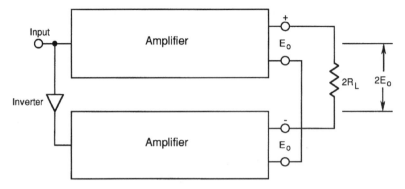

Figure 8-6. Example of amplifier bridging. E_0 = Amplifier output voltage; R_L = Load resistance.

8.6 Amplifier Paralleling

The technique shown in Figure 8-7 has been used in certain sound reinforcement applications where system reliability specifications demand amplifier backup at all times. It is not possible to parallel two amplifiers directly. Any slight imbalance between them is apt to lead to failure of one or both amplifiers. The circuit shown provides isolation between the two amplifiers by means of two autoformers in a balanced bridge circuit and provides *twice* the power into *one-half* the normal load impedance. If one of the amplifiers fails, the output drops by 6 dB, and there will be a signal present at the balance checkpoint, indicating that the system, while still functioning, needs maintenance.

Implementing the technique is expensive if done correctly. Today, this kind of redundancy would be provided by automatic switching.

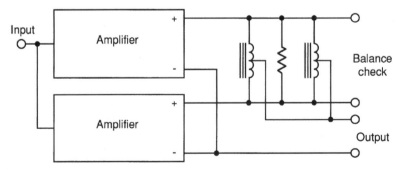

Figure 8-7. Example of amplifier paralleling.

8.7 Biwiring

Many in the audiophile community use what is termed biwiring in connecting the amplifier and loudspeaker. Figure 8-8 shows the basic difference between a normal connection and biwiring. In normal connection, shown in Figure 8-8*a*, a

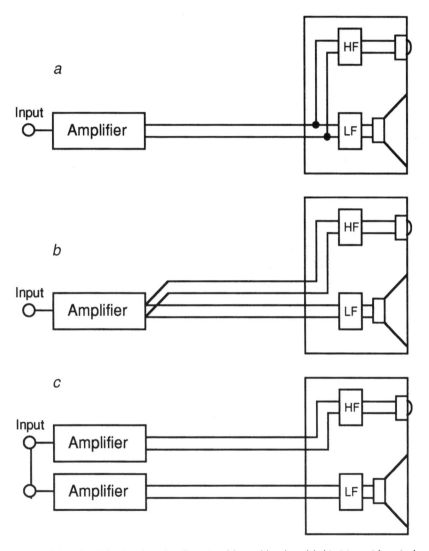

Figure 8-8. Amplifier-loudspeaker biwiring. Normal hookup (*a*); biwiring with a single amplifier (*b*); biwiring with a stereo pair of amplifiers (*c*).

single pair of wires from the amplifier is connected to the two terminals of the loudspeaker, and frequency division takes place in the loudspeaker system. In biwiring, it is necessary for the HF and LF sections of the loudspeaker system to be separately available, each with its full dividing network complement. Each section is then fed from the amplifier through a separate pair of wires, as shown in Figure 8-8*b*.

It is difficult to see what advantages, if any, this offers compared to a single pair of wires of equivalent gauge. However, a variant of biwiring, as shown in Figure 8-8*c*, does offer substantial advantages. Here, the single amplifier has been replaced by two amplifiers, both fed exactly the same full-range program signal. One amplifier drives the LF section and the other drives the HF section. Since LF and HF drives are completely independent, there can be no intermodulation between the two sections of the loudspeaker, and much cleaner sound will result.

8.8 Multiamping

It is just a step away to full multiamping of the system, as shown in Figure 8-9. Here, there is no need for passive crossover components since the necessary

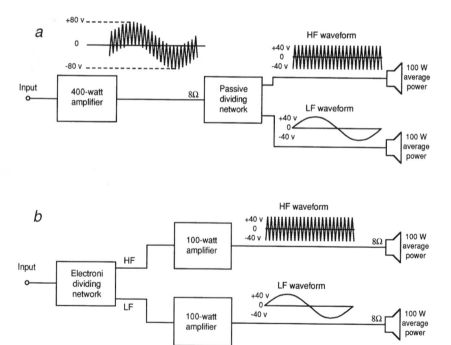

Figure 8-9. Principle of multiamplification. Normal operation (*a*); biamplification (*b*).

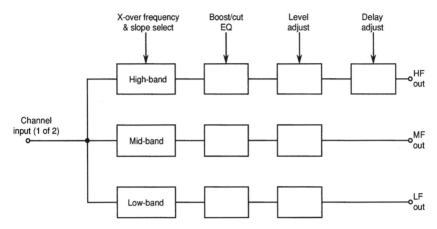

Figure 8-10. Electronic frequency dividing and basic signal processing.

frequency shaping for the system has been carried out ahead of the power amplifiers. In this example it can be seen that a pair of 100-W amplifiers can handle an input signal that would require 400-W capability in a nonbiamplified configuration.

Biamplification is quite common, but triamplification and quadamplification have also been used in larger systems.

8.8.1 Details of Electronic Dividing Networks

Figure 8-10 shows a signal flow diagram for one channel of a multi-amplification frontend. Many such electronic dividing networks are commonly available. Until fairly recently, these have been analog designs. As generic devices, they have adjustable crossover frequencies and slopes, and of course individual output level controls. Many designs include some degree of loudspeaker power response equalization.

Figure 8-11 shows a simpler design for driving a single LF subwoofer in conjunction with a stereo pair of full-range loudspeakers. Here, the passive LF rolloff of the left and right channels feeds the stereo amplifier directly. The two channels are summed, low-passed, and equalized as required for proper LF system operation.

Today, there are many digital controllers for loudspeaker application. These provide the functions of frequency division, time correction, and additional equalization. Such a system is shown in Figure 8-12.

8.9 Electronic Control of Loudspeaker System Performance

Traditionally, one of the great promises in audio engineering has been the prospect of improving loudspeaker performance through the analysis of distortion compo-

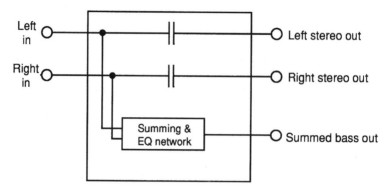

Figure 8-11.   Electronic dividing network for subwoofer operation.

nents and subsequently correcting them electrically. Negative feedback is widely used in electronics for just this purpose, so its application to loudspeakers seems appropriate.

Electronic control of loudspeaker performance is an idea whose time has never really come, perhaps because each step in implementation has been followed by substantial improvements in driver linearity and power handling capability. Most engineers would agree that there is little to be gained by electronically improving the linearity of an ensemble of mediocre drivers. The complexity of the solution might just as well be put into the engineering of better drivers in the first place, through attention to mechanical matters and magnetic circuit design.

Nevertheless, various feedback methods have been used, many to very good effect. Figure 8-13 shows the normal approach that is taken. Here, the velocity or displacement of the cone is monitored, and that signal is reintroduced into the input in opposite polarity. When this is carefully done the system will be stable, and performance will be improved. However, fundamental mechanical limits in the driver cannot be exceeded, so careful limiting of the output signal is called for.

In recent years, as digital processing has become more cost effective, methods of correcting the loudspeaker-room-listener interface have become very popular. Here, the basic plan is to take the room "out of the picture" at lower frequencies, thus ridding the electroacoustical transmission path of much of its normal coloration. The technique shown in Figure 8-14.

Basically, the loudspeaker-listener paths are analyzed at the target listening position via impulse functions fed to the loudspeakers. The impulse response is measured; it is then inverted and used for prefiltering the program material. When carefully done the results can be striking. The listener, however, is fairly limited to a given seating position, and extreme head motions may defeat the correction signal to some extent.

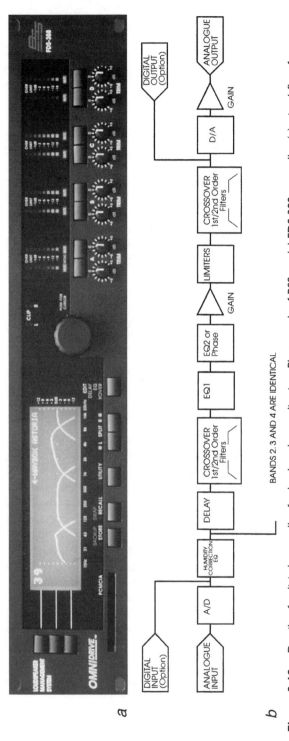

Figure 8-12. Details of a digital controller for loudspeaker application. Photograph of BSS model FDS-388 controller (*a*); signal flow for one frequency band of a stereo pair of channels (*b*). (Data courtesy BSS.)

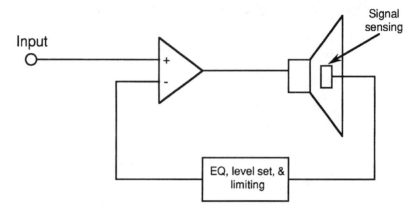

Figure 8-13. Principle of motional feedback applied to loudspeakers.

8-10 Means of Loudspeaker Protection

While a modern amplifier routinely protects itself from adverse loading, loudspeakers themselves have no inherent means of doing this, except perhaps the normal increase in dc resistance that the loudspeaker will exhibit as a result of continuous input at high powers.

One of the simplest methods used in loudspeaker protection is an in-line fuse. Generally, the problem with fuses is that their normal (cool) resistance may high enough to swamp out the beneficial effects of expensive wiring. However, if fuses can be located within the dividing networks, their resistance is normally of little or no concern. The fuse itself, in the power range near burnout, may become nonlinear in its resistance, and this is as matter of further concern. Finally, there is the continuing nuisance of having to replace a blown fuse.

The varistor is a resistance element whose resistance has a positive temperature coefficient; that is, the resistance increases as the temperature rises. One of the best varistors for loudspeaker protection is the ordinary automotive light bulb, properly chosen for the particular current demands at hand. Actually, the light from the bulb, if it is visible through the faceplate of the system's dividing network, can serve as a warning to the operator that trouble may be ahead.

Circuit breakers and shunt Zener diodes are also useful for HF drivers, since they recover quickly and require no maintenance. See Chapter 7 for a discussion of these devices.

In modern system design, such methods and devices as discussed here are fast becoming obsolete. The current crop of electronic dividing networks, whether analog or digital, routinely provide for adjustable limiting thresholds for component protection.

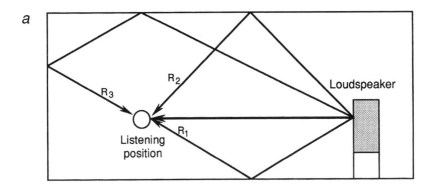

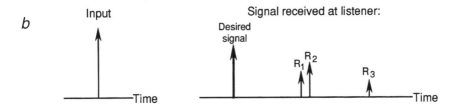

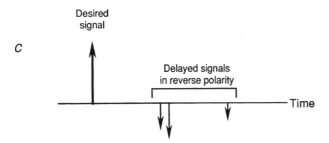

Figure 8-14. Application of digital signal processing for loudspeaker-room enhancement.

Bibliography

Collums, M., *High Performance Loudspeakers*, 4th ed., Wiley, New York (1991).

Giddings, P., *Audio Systems Design and Installation*, Sams, Indianapolis, IN (1990).

Otala, M., and Huttunen, P., "Peak Current Requirements of Commercial Loudspeaker Systems," *J. Audio Engineering Society*, Vol. 35, No. 6 (1987).

CHAPTER

9

Thermal Failure Modes
of Loudspeakers

9.1 Introduction

In professional applications such as sound reinforcement the dominant failure mode of a loudspeaker driver is due to excess heating of the voice coil. Often there is a chain of events in which heating causes expansion of the voice coil, with subsequent rubbing of the voice coil against the top plate and electrical shorting of the voice coil windings.

When we realize that even the sturdiest LF drivers may be no more than 5 or 6% efficient, it is clear that the vast bulk of electrical energy delivered to the driver over time must be converted directly to heat. It is no wonder then that much transducer engineering deals with problems in heat transfer, as well as selection and development of materials and adhesives that are heat resistant.

In this chapter we will outline the basics of heat transfer as they apply to drivers and describe the many measures that have been taken to alleviate the problems. We will also examine the gradual change in driver performance parameters that take place during these processes.

9.2 Basic Heat Transfer Mechanisms

We must have an intuitive understanding of four dimensional quantities: temperature, thermal energy, thermal conduction, and specific heat. Superficially, temperature is the measure of how hot or cold something is; we can sense it by touch. We measure it in degrees Celsius or degrees Fahrenheit, and we can think of it as an *intensive* measure of heat. Thermal energy is the corresponding *extensive* quantity; we simply call it heat, and it is measured in joules. It is temperature that drives heat from one point to another in a thermodynamic system.

The basic notion is shown in Figure 9-1a. Here we have a rectangular bar of material, and we are applying heat (Q) to one end of it. As the bar takes on heat

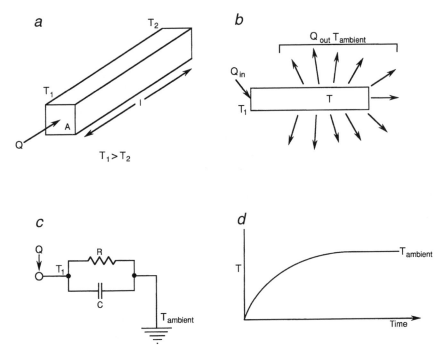

Figure 9-1. Basic heat transfer. Heat applied to the end of a rectangular bar (*a*); at equilibrium, heat introduced into bar equals heat emitted by bar (*b*); equivalent thermal circuit (*c*); temperature rise with time (*d*).

its temperature rises, and heat begins to flow toward the other end of the bar, which is assumed to be at the ambient temperature of the space around the bar.

Let us assume further that heat is being added to the bar at a given rate of some fixed number of joules per second (watts). This, of course, is power, and is the quantity that we normally calculate from measurements of voltage and current. As the bar heats up it it will begin to give off some of that heat to its surroundings by radiation, convection (cooling by the motion of air adjacent to the bar), or by direct conduction to some other object. Eventually the system will come to equilibrium, at which the rate at which heat is applied will be equal to the rate at which it leaves the bar, as shown in Figure 9-1*b*.

While this is in accordance with our everyday understanding of physics, it is a fairly complex phenomenon. First, there is the basic thermal conduction of the bar material itself. Metals such as aluminum and copper have relatively high coefficients of thermal conduction, while other materials such as brick, glass, or concrete have thermal conduction coefficients that are about one-hundredth that of most metals. Accordingly they are known as good thermal insulators.

It may be more convenient to think in terms of thermal resistance than conductance, and the following equation applies to heat transfer along the bar shown in Figure 9-1:

$$R_{\text{thermal}} = l/AK \qquad (9.1)$$

where R is the thermal resistance (°C/W), l is the length of the bar (m), A is the cross-sectional area of the bar (m³), and K is the thermal conductivity (°C-m/w).

There is another quantity, *specific heat*, or heat capacity, which we will now define. Materials vary in the amount of heat required to raise or lower their temperature by some given amount. Specific heat is measured in J°C/kg, the number of joules of energy, per kilogram of material, required to raise the temperature 1°C. We may think of this as thermal capacitance, as given by the following equation:

$$C = mH_s \qquad (9.2)$$

where C is the thermal capacitance (joules × °C), m is the mass (kg), and H_s is the specific heat (J°C/kg).

Over normal operating temperatures, aluminum has a high specific heat—about 20 times that of copper. That is, with equal sample masses, it takes 20 times the amount of heat to raise the temperature of aluminum a given amount, as compared with copper. It is no surprise then that aluminum, because of its high thermal conduction and high specific heat, is the material of choice for heat sinking in electronics manufacturing.

Considering the effects of thermal resistance and specific heat, we can now model the heat transfer process as shown in Figure 9-1c. A typical temperature versus time plot is shown in Figure 9-1d. The equation that defines the rise in temperature is:

$$\Delta T = CR(1 - e^{t/mH_sR}) \qquad (9.3)$$

The thermal time constant, τ, is the time interval during which the temperature has risen to 63% of its maximum value and is given by:

$$\tau = mH_sR \qquad (9.4)$$

9.2.1 Heat Transfer in the Driver

The principles we have just discussed can be used to model the performance of typical cone driver, as shown in the equivalent circuit of Figure 9-2a (Button, 1994). Here, we use two network sections, one representing the relatively short thermal time constant of the voice coil and the other representing the relatively long thermal time constant of the metal structure of the motor.

When we apply a constant power signal to the driver, the voice coil heats up very rapidly to the value ΔT_1 because of its low mass and low conduction path

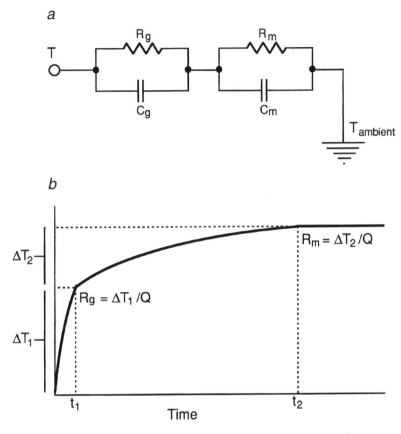

Figure 9-2. Heat transfer in a typical driver. Equivalent thermal circuit for a voice coil and associated metal structure (*a*); temperature rise with time (*b*).

to the adjacent metal. After that state has been reached, we then observe a much longer time constant as the temperature further attains the added value of ΔT_2 as heat is conducted away from the voice coil through the metal structure to the outside ambient temperature.

The total temperature rise will be $\Delta T_1 + \Delta T_2$, and the plot of overall temperature rise with time is shown in Figure 9-2*b*.

9.3 Estimating Values of Thermal Resistance

There is no direct way to measure thermal resistances in a complex structure such as a dynamic transducer, and indirect methods are used. Button (1994) estimated values of R_g and R_m by applying signals to the driver and closely tracking the rise in voice coil resistance. Various voice coil diameters and lengths were measured, as were various magnetic structures. Two types of excitation signals were used, LF noise and a MF sine wave. Button's data can be summarized as follows:

1. With LF noise input, the value of R_g was reduced. This result indicates that significant voice coil motion itself aids in heat removal, either through turbulence or through closer proximity of all parts of the voice coil to the top plate and pole piece.

2. Increasing input power reduces R_g. Increased turbulence is a factor here, as is the expansion of the voice coil, placing it closer to the top plate where heat conduction is increased.

3. All other factors remaining equal, larger voice coils run cooler than smaller ones, due to their increased surface area.

4. More massive voice coils exhibit slower heating and as such will have better thermal transient capability. This suggests possible design trade-offs in that lighter voice coils have lower mass and inductance, both of which improve transient response.

5. Since most drivers have roughly the same aluminum frame volume and surface areas, values of R_m were virtually the same. (One significant departure was observed with a driver configuration in which thermal short-circuiting of the voice coil to the frame was apparent.)

6. Excessive voice coil overhang may be detrimental. Even with high-level noise signals, the outer portions of the voice coil remain far from the top plate and pole piece and thus cannot take advantage of that low-resistance thermal path.

9.4 Low Frequency Performance Shifts

Figure 9-3 shows the on-axis response of a 380-mm driver with inputs varying from 0.8 W to 100 W in 3-dB increments, all with the same vertical scale. Examined casually, it appears that subsequent increments are in fact 3 dB. A clearer way to see what is really going on is to plot, say, 1-W and 100-W curves with a 20-dB offset between them, as shown in Figure 9-4. Here, we can clearly see that the 100-W curve is compressed approximately 2 dB relative to the 1-W curve. The effect is often referred to as *dynamic compression*.

One may question whether this degree of compression is audible as such. The answer to this question may be seen in the data shown in Figure 9-5. The frequency and impedance response of a 460-mm LF driver at normal temperature (27°C) are shown in Figure 9-5a. The response at an elevated temperature of 150°C is shown in Figure 9-5b. Note that the LF alignment has been shifted by a considerable amount which would be quite audible to experienced listeners.

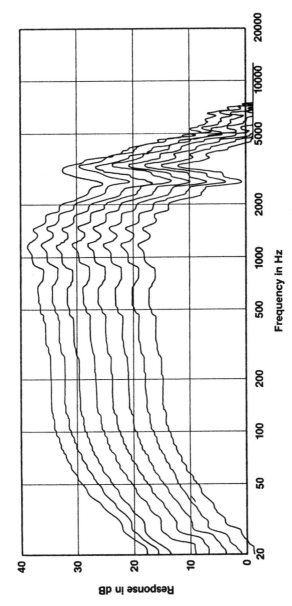

Figure 9-3. Axial response of a 380-mm LF driver at 1 m driven at input powers of 0.8, 1.6, 3.15, 6.3, 12.5, 25, 50, and 100 W. Bottom line is 80 dB Lp. (Data courtesy JBL)

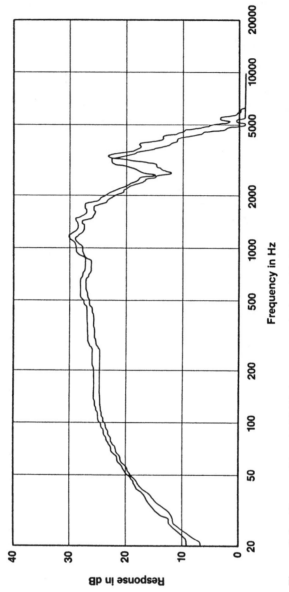

Figure 9-4. Curves of 1 and 100 W, 1 m on-axis, displaced 20 dB. Bottom line is 70 dB Lp for 1 W and 90 dB Lp for 100 W. (Data courtesy JBL, Inc.)

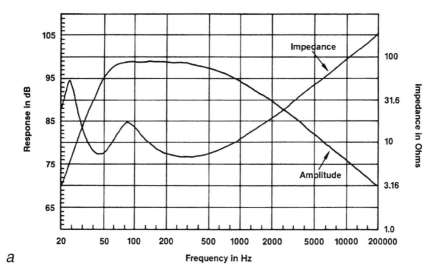

a

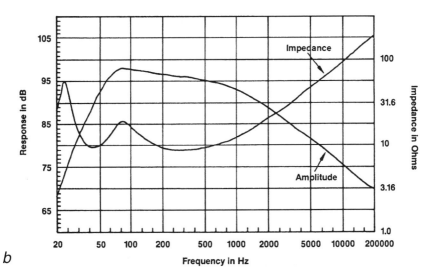

b

Figure 9-5. Low-frequency alignment shifts. Normal operation at 27°C (*a*); elevated operation at 150°C (*b*). (Data courtesy JBL, Inc.)

9.5 Techniques for Heat Removal

Air convection and turbulence are very effective methods of removing heat from the magnetic gaps of large drivers. By directly opening up portions of the gap area to the outside of the driver, JBL's Vented Gap Cooling (VGC™) provides a set of paths through which air can be pumped in and out at high excursions.

The effectiveness of the VGC design is shown in Figure 9-6. Here, the on-axis output versus time is shown for three 380-mm LF drivers. The VGC design is compared with a standard design, as well as with a smaller voice coil diameter design.

9.5.1 Ferrofluids

Many smaller cone and dome drivers used for MF or HF applications are often designed with ferrofluids in their magnetic gaps. Ferrofluids consist of a suspension of extremely fine iron oxide particles in a liquid mixture that functions as a lubricant, damping substance, and a surface active agent that prevents the magnetic particles from clumping together. The material is applied to the magnetic gap with the voice coil in place.

Without such an agent, the voice coil of the typical 25-mm dome HF device will heat up up very rapidly due to its very small size and the poor thermal conductivity provided by the air path to the metal portions of the driver. Ferrofluid

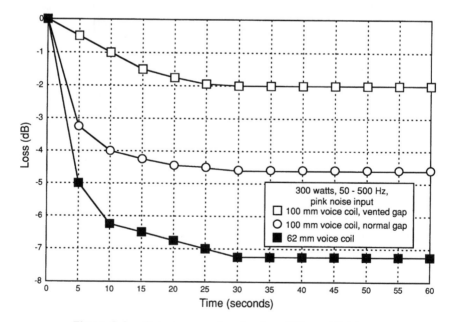

Figure 9-6. Output versus time for three 380-mm LF drivers.

can effectively short circuit the thermal path to the metal structure, giving the driver a much longer thermal time constant and allowing it to handle relatively high electrical transient input signals.

If the viscosity of the ferrofluid is properly chosen, the HF sensitivity of the driver will be only slightly diminished. An added benefit in some designs is that ferrofluid can be used to damp the fundamental resonance of the driver. There may also be a sight advantage in reducing the reluctance of the magnetic path, resulting in an increase in output of perhaps 0.75 dB, depending on other driver parameters.

Ferrofluids may be used in LF drivers, but are generally not used in those drivers intended for large excursions, inasmuch as nonlaminar flow may result, with consequent nonlinearities.

Early ferrofluids presented problems of aging and stability under high-temperature drive conditions, but these problems have largely been solved in recent years.

Bibliography

Collums, M., *High Performance Loudspeakers*, John Wiley & Sons, New York (1991).

Button, D., "Heat Dissipation and Power Compression in Loudspeakers," *J. Audio Engineering Society*, Vol. 40, No. 1/2 (1992).

Button, D., "Heat Dissipation and Power Compression in Low Frequency Transducers," *Sound & Video Contractor*, Vol. 12, No. 2 (February 1994).

Henricksen, C., "Heat Transfer Mechanisms in Loudspeakers: Analysis, Measurement, and Design," *J. Audio Engineering Society*, Vol. 35, No. 10 (1987).

Mellilo, L., and Raj, K., "Ferro-fluid as a Means of Controlling Woofer Design Parameters," *J. Audio Engineering Society*, Vol. 29, No. 3 (1981).

Recording Monitor Loudspeakers

CHAPTER

10

10.1 Introduction

The term "monitor loudspeaker" is loosely applied to almost any loudspeaker used for monitoring recorded product at any stage in sound production or postproduction. In a more restrictive sense the term applies to products that are widely accepted in monitoring operations and have, to some extent, been modified or designed to meet a list of performance attributes determined by recording and broadcast engineers. Broadly speaking, those attributes are:

1. *Extended bandwidth.* The range from 40 Hz to 16 kHz ($\pm$ 3 dB) represents the minimum acceptable for a full-size monitor. Smaller monitors should be just as flat across their passband, but the allowable LF limit may be raised somewhat.

2. *Flat frontal angle response.* The bandwidth stated in item 1 should be uniform over a horizontal beamwidth of $\pm$ 15° and a vertical beamwidth of $\pm$ 5°, both with respect to the forward axis.

3. *Controlled power response.* The DI of the system should be free of any deviations exceeding ± 3 dB over the range from 250 Hz to 10 kHz.

4. *Accurate time domain response.* The group delay of the system should fall within the Blauert and Laws criteria (see Section 5.7).

5. *Accurate stereophonic imaging.* If the loudspeaker is not inherently of horizontal mirror image symmetry, it should be offered in separate left and right models so that symmetrical listening geometry can be achieved.

6. *Robust construction with high reliability.* In the most general terms, the system should be well constructed, roadworthy, and able to handle its share of inadvertent abuse.

7. *Well behaved impedance characteristic.* The dc resistance of the system should not drop below 80% of the nominal impedance value, and the phase angle should not exceed $\pm 60°$.

200

8. *Low distortion at normally required operating levels.* The system should be able to handle the levels expected of it with little, if any, audible distortion. In the rock studio, level requirements for a stereo pair of monitors at a distance of 3 m, may be in the range of 115 dB Lp.

This is an imposing list, and few monitor loudspeakers earn outstanding marks in all areas. In many cases, the optimization of one attribute may work against another. Typically, a monitor system optimized for classical music recording, with its demands of very low distortion and flat frequency response, may not be up to the high volume level demands of a modern pop/rock studio. And conversely, the typical HF horn system used in the pop/rock studio may have a much too aggressive sound for the classical engineer and producer.

In this chapter we will discuss monitor attributes in detail and present examples of well-known designs that have met most of these requirements.

10.2 A Historical Survey

As monitor systems developed during the thirties and forties, a premium was placed on efficiency, inasmuch as power amplifiers of that era generally offered no more than 20- to 40-W output capability. The normal approach was to use a high-efficiency LF driver, coupled with a HF horn/compression driver. The earliest such system is unquestionably the Iconic, first manufactured in the thirties by the the Lansing Manufacturing Company. The utility form of the product is shown in Figure 10-1. This was a two-way system, as are many modern systems with HF horns. The LF driver had a rather high resonance frequency for increased efficiency, and the HF portion consisted of a compression driver/multicellular horn combination, crossing over in the range of 800–1200 Hz. Because high-energy magnets were not available at the time, dc field coils were used in both LF and HF transducers to generate the required magnetic flux densities. The field coil power supply (with vacuum tubes) can be seen on the top of the enclosure. By today's standards, such a system had limited bandwidth and irregular frequency response—but doubtless excelled in the areas of ruggedness and high output capability.

The forties saw the rise of several so-called coaxial two-way designs, in which the HF unit is located in the center of the LF cone. Most notable of the U.S. products was the 380-mm (15 in.) Altec-Lansing 604, which is still made today. So widely has the 604 been used over the years that during the seventies several companies designed their own dividing networks for the basic unit. The Mastering Lab network was in wide use for several years, and the UREI network was first employed with the basic Altec-Lansing unit plus one or more auxiliary LF drivers. UREI was subsequently acquired by JBL, and the UREI systems were adapted to a composite driver made by JBL. Figure 10-2*a* shows a side/half-section view

Figure 10-1. Photograph of Lansing Iconic monitoring system. (Data courtesy JBL, Inc.).

of the Altec 604, and Figure 10-2*b* shows a similar view of the UREI driver. Both designs are still made today.

Another noted U.S. loudspeaker was the RCA LC-1A coaxial model. This was a 380-mm (15 in.) LF driver with a uniquely stiffened LF cone and a small HF cone located at its apex. A drawing of this system is shown in Figure 10-3.

In England, the Tannoy Dual-Concentric design followed the general form and size of the 604, but with the unusual distinction of using its curvilinear LF cone as an extension of the HF horn. A section view is shown in Figure 10-4.

10.3 The Modern Era

Since the eighties, monitor loudspeakers have taken two distinct paths. There are those adherents to the traditional horn HF approach, with normally ported

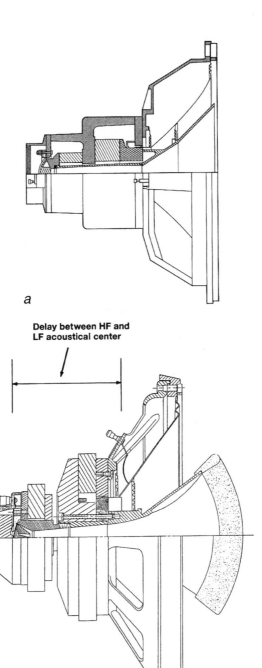

a

**Delay between HF and
LF acoustical center**

b

Figure 10-2. Side/half-section view of Altec 604 380-mm driver (*a*); side/half-section view of UREI composite 380-mm driver (*b*). (Data courtesy Altec Corp. and JBL, Inc.).

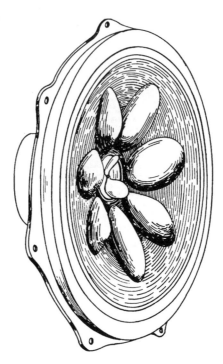

Figure 10-3. View of RCA LC-1A.

LF sections. But many engineers and producers are now finding that the newer cone/dome monitors, if outfitted with multi-amplification and judicious signal limiting, can easily handle high studio playback levels.

The primary differences between horn HF systems and cone/dome systems is that the former produce greater second harmonic distortion at high frequencies at normal operating levels (95–110 dB Lp at 1 m) than do the cone/dome systems. On the other hand, the cone/dome system ordinarily has to be operated as a multiway system, with its inevitable problems in lobing and driver interference. The horn HF system can easily be designed for two-way operation, with minimal lobing problems. Finally, there is the ability of the horn to withstand all manner of assaults from the driving amplifier and still hold together, albeit with significant distortion.

The bottom line, then, is dependent on how loud the engineer and producer wish to monitor. For moderate levels, there appears to be an advantage with cone/dome systems where distortion is concerned. For rock recording there will probably always be an advantage to the HF horn approach.

We will now take a closer look at some of these newer systems.

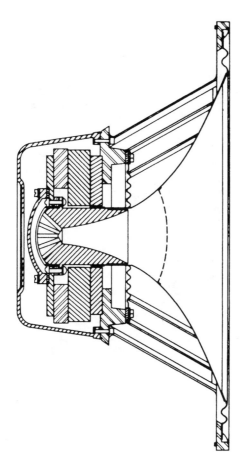

Figure 10-4. Section view of Tannoy 380-mm Dual-concentric driver. (Data courtesy Tannoy.)

10.3.1 A Modern Horn High-Frequency System

The design shown in Figure 10-5*a* is the JBL DSM-1 system. The dual LF units, above and below the HF section, create a d'Appolito array, and this approach has become very popular in recording monitoring applications in recent years. The system is digitally controlled, and the net drive curves for producing flat on-axis response are shown in Figure 10-5*b*. Acoustical output and distortion are shown in Figure 10-5*c*.

The rise in drive voltage above about 3 kHz is approximately the inverse of the fall off in HF response in compression drivers that we discussed in Chapter 7. The HF system makes use of rapid flare rates in both driver and horn for lower second-harmonic distortion (see Section 7.5).

a

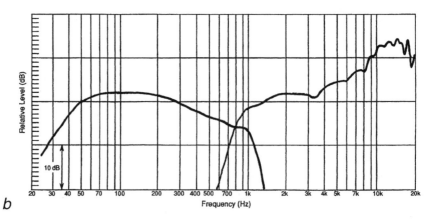

b

Figure 10-5. The JBL DMS-1 Monitor. Photograph of unit (*a*); LF and HF drive voltage curves (*b*); on-axis response and distortion (+20dB) response (*c*). (Data courtesy JBL, Inc.).

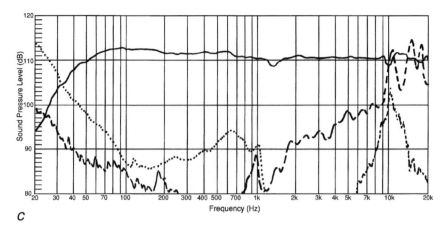

C

Figure 10-5. *Continued*

10.3.2 A Modern Cone/Dome System

Figure 10-6a shows baffle details of a Genelec Model 1038A monitor system. The design is three-way, triamplified, with each amplifier section providing the necessary frequency division and limiting. User-adjustable equalization provides for various boundary effects and helps in interfacing the system with the listening environment. A signal flow diagram is shown in Figure 10-6b and system response is shown at c.

Figure 10-7 shows a moderate-size three-way cone/dome system made by Westlake Audio intended for use in Musical Instrument Digital Interface (MIDI), postproduction, and remote operations.

In addition to cone-dome loudspeakers specifically designed for monitoring, a handful of consumer systems have been widely used by classical music recording engineers as reference systems. Notable here is the B&W model 801.

10.3.3 Thermodynamic Distortion in Dome Systems

Figure 10-8 shows the thermodynamically induced distortion produced by a 25-mm (1-in.) dome on sine wave input signals. As a general rule we can state that the sustained sine wave input to a 25-mm device is about 20 W. The normal maximum sensitivity of these devices is limited to about 93 dB, 1 W at 1 m. Therefore, a single unit driven at 20 W will produce a level of 106 dB at a distance of 1 m. From the graph we can see that the resulting thermodynamic distortion will be between 1 and 2%, a fairly low figure.

For comparison with a horn HF system, we can see in Figure 10-5c, that the second harmonic distortion for a level of 100 dB Lp at 1 m is in the 3% range above 10 kHz. We hasten to remind the reader at this point that with 20 W input

a

Figure 10-6. The Genelec Model 1038A Monitor. Photograph of unit (*a*); signal flow diagram (*b*); response (*c*). (Data courtesy Genelec.)

we have already reached the power output limit of the 25-mm dome. At these same levels, the compression driver has considerable headroom available.

10.4 Monitoring Environments

There is always a strong argument for monitoring a recording in an environment that resembles a typical living room. Indeed, the IEC has proposed a "standard room" for making loudspeaker subjective listening tests that is, in essence, the average of a number of typical listening spaces. While a classical engineer may prefer the comfort of a living room environment, the pop/rock recording engineer generally prefers the conditions that exist in a typical control room. In the control room the primary monitor loudspeakers are usually soffit mounted; that is, they

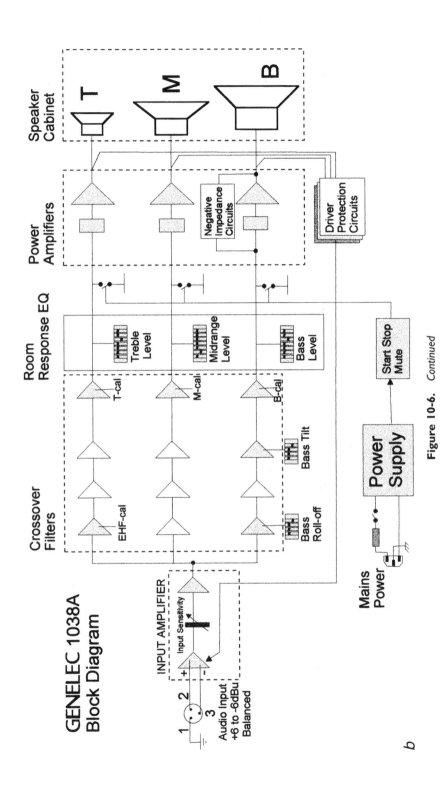

Figure 10-6. *Continued*

209

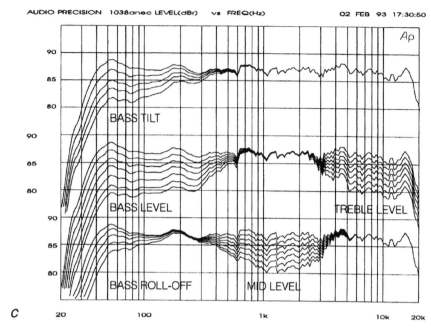

Figure 10-6. *Continued*

Figure 10-7. Westlake Audio Model BBSM-8 monitor. (Data courtesy Westlake Audio.)

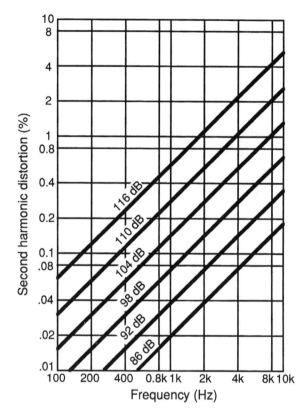

Figure 10-8. Twenty-five-millimeter (1 in.) dome thermodynamic distortion as a function of frequency and level at a distance of 1 m. (Data after Thuras et al., 1935.)

are flushed into the architecture on a diagonal surface at the junction between wall and ceiling. There are several reasons for this; primarily, it reduces discrete loudspeaker reflections from the large expanse of glass at the front of the control room. Secondly, it conserves space—which there is never enough of in a control room.

Augspurger (1990) points out that the average ratio of direct to reflected sound in a wide variety of monitoring environments is unity at the listening position. That is, there is as much direct sound as reflected sound. The job of the studio designer is to ensure that the reflected sound in the control room is diffuse enough so that it does not interfere with the perception of direct sound. On the other hand, if the reflected sound is too low in level the listener will feel the oppressiveness of such a dead acoustical environment. One way of handling this problem is to make the front portion of the monitoring space fairly absorptive and the back portion fairly live, with good diffusion (Davis and Meeks, 1982).

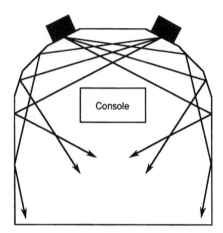

Figure 10-9. A reflection-free zone around a recording console.

Another concern in monitor space design is to minimize direct reflections from nearby surfaces. Figure 10-9 shows the normal arraying of surfaces for this purpose.

Low frequencies present a problem in small monitoring spaces. If the room is not sufficiently damped, discrete room response modes will become apparent, adding midbass coloration to the sound. The easiest way around this problem is to damp all low frequencies by means of so-called bass traps. These are architectural elements that are fairly deep (often on the order of a meter), filled with absorptive materials, and intended to minimize standing waves. Since bass traps absorb LF sound energy, it is usually necessary to reinforce the LF capability of the primary monitoring system to restore the net LF capability of the system. Thus, we might see several 460-mm (18-in.) subwoofers added to the system in order to restore this balance.

10.5 The Near-Field Monitor

The modern control room is not complete without several pairs of loudspeakers used as so-called near-field monitors. These are small systems, normally placed on the meter bridge of the recording console, and are are intended to give the producer and remix engineer a good idea of how the mix will sound on small loudspeakers, such as minisystems in the home and, of course, automobile stereo systems.

While almost any loudspeaker can be used here, the major suppliers of large

Figure 10-10. Photograph of a family of small bookshelf monitors. (Data courtesy JBL Professional, Inc.)

monitor loudspeakers routinely provide small models intended for this purpose. Figure 10-10 shows a family of such loudspeakers.

For the most part, loudspeakers intended for this purpose differ little from home hi-fi models, except in matters of appearance.

10.5.1 The Home Project Studio

During the last decade the lowly home studio has come of age, and many successful recordings are now made in these informal spaces. The driving force has been MIDI control of electronic instruments and the rise of modular digital multitrack (MDM) recorders.

In most cases there is little, if any, acoustical recording done in these spaces. If a vocal line needs to be added, a multitrack work tape can be taken to a full-line studio and the vocal added there. Monitoring in these project studios is relatively simple, and is often done entirely over headphones. Near-field monitors are the best choice in these applications.

10.6 Monitor System Equalization

Where careful attention has been paid to architectural acoustical details and to monitor specification, response of the left and right loudspeakers may be virtually identical, with only slight differences between them. When channel balance is this close, it may be expedient to use a pair of high-quality, 1/3-octave equalizers to fine-tune the systems, so to speak, so that they have the same response at the prime listening position. Using these equalizers for any other purpose may be a mistake.

Figure 10-11a shows details of the instrumentation used in monitor system equalization. The output of a pink noise (equal power per octave) generator is fed to the equalizer-amplifier-loudspeaker chain, and the filter sections carefully adjusted so that the acoustical output, as measured on a real-time analyzer matches the desired contour. The filters used for the purpose are minimum phase, and their response usually provides a direct complement to that of the loudspeaker drivers themselves. The functioning of the real-time analyzer is shown in Figure 10-11b.

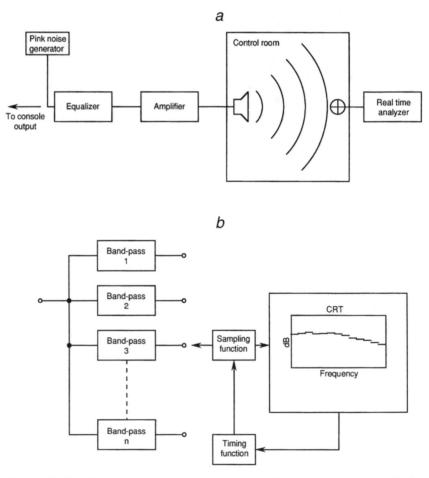

Figure 10-11. Monitor system equalization. A pink noise generator is inserted in the monitor chain just before the equalizer (a); details of the real-time equalizer (b).

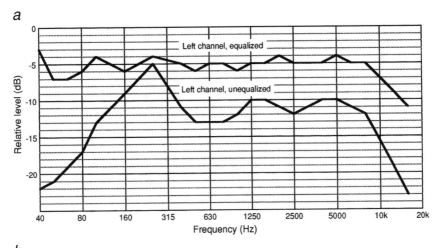

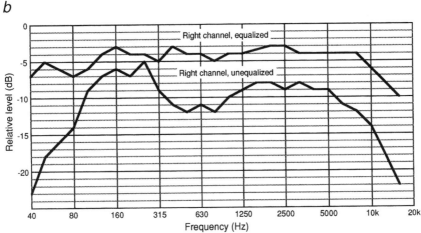

Figure 10-12. An example of monitor equalization. Left channel amplitude response before and after equalization (*a*); right channel amplitude response before and after equalization (*b*); equalizer amplitude responses (*c*); equalizer phase responses (*d*).

A typical application is shown in Figure 10-12*a* and *b*. Here, the unequalized left and right channels are reasonably balanced in the monitoring space. However, due to flexible wall structures, LF absorption is considerable, requiring a significant boost in both channels to maintain flat acoustical response down to the 40 Hz range. A 6 dB per octave rolloff above 8 kHz is evident. The amplitude and phase response of the adjusted equalizers are shown in Figure 10-12*c* and *d*.

Finally, we present a number of target monitor system equalization curves. While it may seem natural to equalize monitor systems for flat response, a survey of home listening environments indicates that normal system response is rolled off at high frequencies, due primarily to the interaction of room acoustics with

c

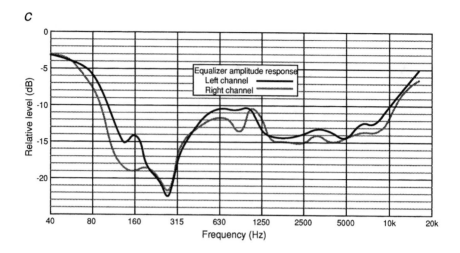

d

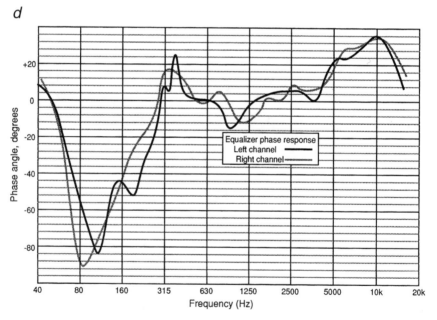

Figure 10-12. *Continued*

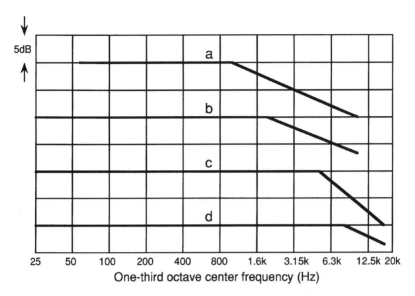

Figure 10-13. Target equalization curves for monitoring. Curve a is used widely in sound reinforcement. Curve b is used in motion picture systems. Curves c and d have been used in recording control room equalization, with a general preference for curve d. Tolerances of ±1.5 dB can generally be met above 200 Hz in control rooms, while tolerances of +2 and −4dB can be met below 200 Hz. In well designed rooms, smoother response can be expected below 200 Hz.

the axial and power response of typical home high-fidelity loudspeakers. If the engineer and producer balance a recording over flat monitors that sounds correct to them, it may sound dull in some listening environments.

This information leads, at least tentatively, to the conclusion that the final mastering/mix-down environment for recordings should represent a center-line value in terms of stereo system equalization in the average home environment. Accordingly, the curves shown in Figure 10-13 provide guidelines for what should be done.

Bibliography

Augspurger, G., "The Importance of Speaker Efficiency," *Electronics World*, January 1962.

Augspurger, G., "Versatile Low-Level Crossover Networks," *db Magazine* (March 1975).

Augspurger, G., "Loudspeakers in Control Rooms and Living Rooms," in *Proceedings of the Audio Engineering Society 8th International Conference*, Washington, DC (1990).

Cooper, J., *Building a Recording Studio*, Recording Institute of America, New York (1978).

Davis, D., and Meeks, G., "History and Development of the Control Room Concept," Preprint No. 1954, Audio Engineering Society Convention, Los Angeles (1982).

Eargle, J., "Equalizing the Monitoring Environment," *J. Audio Engineering Society*, Vol. 21, No. 2 (1973).

Eargle, J., "Requirements for Studio Monitoring," *db Magazine* (February 1979).

Eargle, J., *Handbook of Recording Engineering*, 3rd ed., Chapman & Hall, New York (1995).

Eargle, J., and Engebretson, M., "Survey of Recording Studio Monitoring Problems," *Recording Engineer/Producer*, Vol. 4, No. 3 (1973).

Engebretson, M., "Low Frequency Sound Reproduction," *J. Audio Engineering Society*, Vol. 32, No. 5 (1984).

Fink, D., "Time-Offset & Crossover Design," *J. Audio Engineering Society*, Vol. 28 (1980), No. 9.

Lansing, J., "The Duplex Loudspeaker," *J. Society of Motion Picture Engineers*, Vol. 46 (1946), pp. 212–219.

Smith, D., Keele, D., and Eargle, J., "Improvements in Monitor Loudspeaker Design," *J. Audio Engineering Society*, Vol. 31, No. 6 (1983).

Thuras, et al., *J. Acoustical Society of America*, Vol. 6, No. 3 (1935).

Loudspeakers in
Sound Reinforcement

CHAPTER

11.1 Introduction

Reinforcement of speech and music has long been commonplace, and performance standards are continually on the rise. With care, the all-important aspects of naturalness and system stability can be maintained, and today many major musical events are performed with reinforcement of both orchestra and vocalists.

One reason for this is the use of larger venues for performance. For reasons of return on investment, producers of music and stage events prefer to play to as large a house as possible. Outdoor music events are routinely reinforced. When the Hollywood Bowl was built in the twenties, there was no sound reinforcement. There were also no freeways and few if any overflights. Today, the ambient noise level everywhere is on the rise, and sound reinforcement is a way to compensate for it.

Modern pop music, wherever it is performed, is reinforced, primarily because it nearly always is first heard over loudspeakers and is first distributed by recordings. Generally, young people prefer their music to be loud, often to the detriment of their long-term hearing acuity.

11.2 Systems for Speech Reinforcement

The general requirements for intelligible speech reinforcement are:

1. Good loudspeaker coverage of all patrons.
2. Absolute system stability (freedom from feedback).
3. Adequate signal level at each listener above the ambient noise level in the listening space.
4. Adequate signal to reverberant level at each listener.
5. Suitably short reverberation time in the listening space.
6. Freedom from discrete reflections and echoes in the listening space.

To these requirements we can add three more that will enhance the quality of listening:

7. Naturalness of sound; the spectrum of reinforced sound should reasonably match that of the talker.
8. The perceived direction of reinforced sound should correspond to that of the talker.
9. The reinforced sound should be substantially free of distortion and be able to reach realistic levels.

11.2.1 Central Arrays versus Distributed Systems

Historically, there have been two approaches to speech reinforcement, the central array and the distributed array. Generally, the central array is preferred from the point of view of satisfying the nine requirements stated in the previous section. However, in very live spaces the distributed array is often more cost effective. The choice here can be seen from the decision flow diagram shown in Figure 11-1.

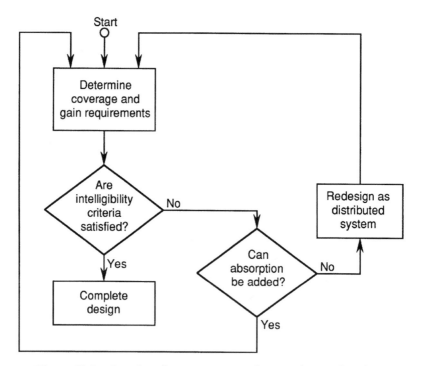

Figure 11-1. Sound reinforcement system design, a decision flowchart.

A central array will work if it can be located so that the longest "throw" of the system will be no more than about four times its height above the floor. There is the further assumption that reverberation time, direct-to-reverberant level, and the pattern of discrete reflections in the listening space satisfy items three to five in the foregoing list of requirements. In highly reverberant spaces it may be impossible to meet these requirements with a central array, and if sufficient absorption cannot be added to the space, the system designer will have to opt for a distributed system.

A distributed system is one in which there are many loudspeakers, each covering only a small portion of the listening area. Since these loudspeakers are located close to the listeners, they will provide a higher direct-to-reverberant ratio at the listener than will the central array, thereby satisfying requirements two and three. However, this is often done to the detriment of perceived naturalness of the reinforced sound.

11.3 The Role of Signal Delay

Through the use of signal delay devices, the performance of distributed arrays can be much improved in terms of naturalness. Signal delay also facilitates the use of both central and distributed arrays in the same space. In general, the use of delay makes it possible to compensate electrically for the acoustical delay that occurs over distance. The principle is shown in Figure 11-2*a*. If a talker is directly amplified and fed to an overhead loudspeaker, the listener will localize the source of amplified sound directly above. If the signal is delayed by an amount equal to $x/344$ s, where x is the talker-listener distance in meters, then both direct and amplified sound will arrive at the listener at the same time. Because the overhead source will be louder, the listener will still localize the source directly above. However, if the signal is further delayed by a few milliseconds, the source of sound will appear to be from the front.

The trading value between level and signal delay is indicated in Figure 11-2*b*. For a given excess signal delay to the overhead loudspeaker, as indicated on the horizontal scale, the overhead loudspeaker can be driven at a higher level, as shown, with localization tending toward the front.

Another important point in system design is shown in Figure 11-3. Whatever the design approach, it is essential that reinforced sound be aimed where it will do the most good—at the listener. Acoustical power radiated indiscriminately, as shown in Figure 11-3*a*, will result in a greater reverberant level in the listening space than if that same power is radiated into the fairly absorptive audience area, as shown in Figure 11-3*b*. The absorption coefficient of an audience at mid- and high frequencies is in the range of 0.8, indicating that the initial reflected sound level will be 20%, or 7 dB lower than the impinging sound.

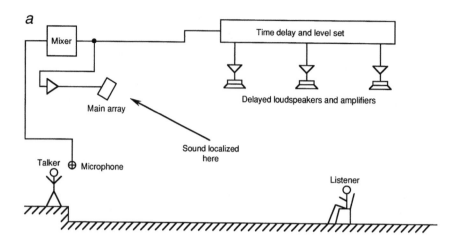

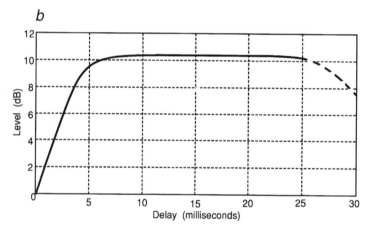

Figure 11-2. The Haas, or precedence, effect. The listener will localize the source of sound at the first arrival, even though a later arrival may be significantly louder (*a*); range of precedence effect; beyond about 25 ms, the listener begins to hear the delayed sound as a discrete echo (*b*).

11.4 Case Studies

A few case studies will show how these techniques are normally integrated into sound systems.

11.4.1 Central Array with Delayed Underbalcony Loudspeakers

The system shown in Figure 11-4 is typical of most auditoriums. The central array is located above the proscenium in the center of the house. Horns are

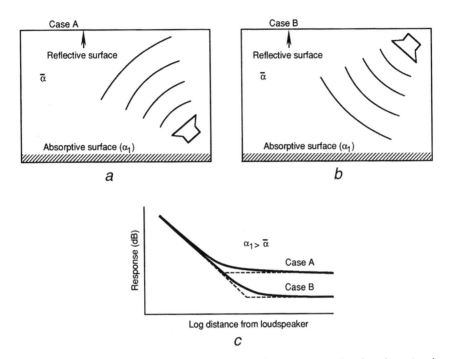

Figure 11-3. Controlling reflected sounds in a performance space. Loudspeakers aimed at reflective surfaces produce a relatively high reverberant field level (*a*); loudspeakers aimed at the highly absorptive audience area will reduce the reverberant field level proportionally (*b*); attenuation of sound with distance under the conditions in both (*a*) and (*b*) (*c*).

chosen for their vertical and horizontal nominal coverage angles and splayed along their −6-dB zones in order to provide coverage at mid- and high-frequency for all parts of the seating area that are visible from the array. It is advantageous for the horn mouths to be adjacent to each other. If the horns are not of the same length, the signal to the shorter horns may be delayed so that their electrical acoustic centers are substantially in alignment with the longer ones, as shown.

The underbalcony array of loudspeakers is then delayed so that the sound arriving at the listener is slightly delayed, relative to the spillover sound from central array. In this manner, the listeners well under the balcony will be able to hear clearly, and those listeners in the transition zone at the front of the underbalcony area will not be bothered by disturbing echoes produced by the dual sources.

The loudspeakers used in the design of the central array are chosen for their precise coverage angles (see Section 7.4.5). Those mounted in the balcony soffit are chosen for the broadest hemispherical coverage over the frequency range up to about 2 kHz. Nominally, the target angle for −6 dB coverage is 90°.

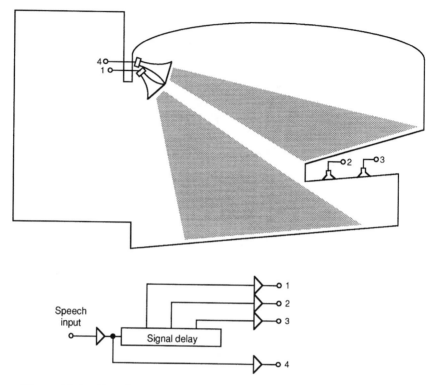

Figure 11-4. Use of signal delay in a large theater sound reinforcement system.

11.4.2 A Pewback System

The pewback system is often used in large, reverberant worship spaces. Essentially, it is a way of reducing the loudspeaker-listener distance so that the listener is clearly in the direct sound field of the nearest loudspeaker and effectively isolated from the acoustics of the room. Details are shown in Figure 11-5.

Normally, there is a target loudspeaker at the front of the space. The distributed loudspeakers are located every 1.5 m or so on the backs of the pews and are zoned for signal delay every 6 m (20 ms) as measured from the target loudspeaker at the front so that the first arrival sound at each listener will be from the front of the room. The delay interval of 20 ms is chosen so as not to exceed the range over which the Haas effect will provide clear localization without echoes. In this way, the listener will tend to localize sound from the front of the space, even though the bulk of the sound heard by the listener is actually coming from the nearest pewback loudspeaker. The loudspeakers chosen for this purpose are small models of perhaps 75 mm diameter with fairly wide frequency range. Whizzer cones are commonly used for their good HF dispersion. The LF bandwidth may be limited to 125 Hz.

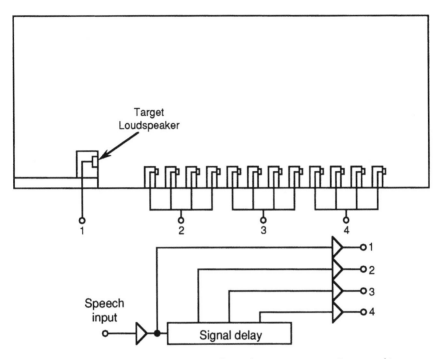

Figure 11-5. Details of a pewback speech reinforcement system in a worship space.

These systems are normally used only with speech. Music and other liturgical activities are not generally reinforced over these systems, and rely on the natural acoustics of the space.

11.4.3 Speech Reinforcement in Large Public Spaces

Modern transportation terminals and convention facilities have large spaces of extended area with relatively low ceilings. Central arrays are normally out of the question here, and ceiling mounted loudspeakers are the usual choice, primarily for good coverage as well as the ease with which they can be zoned for local announcements. When these systems do not work well, it is normally due to insufficient loudspeaker density in the ceiling layout.

The layout density determines the smoothness of coverage. No loudspeaker should be used in distributed systems if it does not have a nominal coverage angle of 90° (−6 dB) at 2 kHz. Two loudspeaker layout patterns are commonly used, triangular and square; these are shown in Figures 11-6 and 11-7, respectively. Three versions of each layout are shown: edge-to-edge, minimum overlap, and 100% overlap. The recommended choice with either configuration are minimum and 100% overlap. The response variation for each configuration over the coverage area is indicated in Figures 11-6*d* and 11-7*d*.

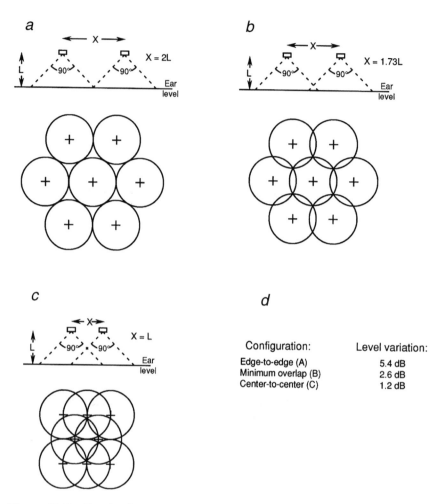

Figure 11-6. Details of triangular ceiling loudspeaker layout. Edge-to-edge layout (*a*); minimum overlap (*b*); 100% overlap (*c*); expected level variations at the ear plane (*d*).

Large meeting spaces are often provided with distributed systems. In these cases there might be demands for fairly high music levels, and the ceiling requirement may call for a large number of 380-mm (15-in.)-diameter drivers mounted in suitable enclosures. A matter often forgotten here is the deflection of the driver's cone due to gravity when mounted vertically. This will result in nonlinear performance which may be aggravated over time. The amount of cone deflection is inversely proportional to the square of the free-air resonance frequency of the driver and is shown in Figure 11-8. As high a driver resonance frequency, consistent with the required signal bandwidth, should be chosen for these applications.

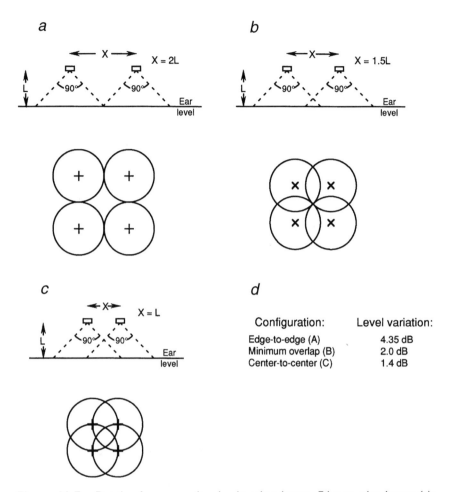

Figure 11-7. Details of square ceiling loudspeaker layout. Edge-to-edge layout (*a*); minimum overlap (*b*); 100% overlap (*c*); expected level variations at the ear plane (*d*).

11.5 Computer Simulation of Loudspeaker Coverage

There a several personal computer programs available for systems designers and consultants in laying out loudspeaker systems in large spaces. The user first enters room boundary and finishing data. Then one or more loudspeakers are chosen, mounted, and oriented in the space. The program then computes the values of direct field coverage and a variety of other readouts, including direct-to-reverberant ratio and an estimate of system intelligibility. Various loudspeaker pattern merging strategies may be used.

Another part of some of these programs provides a level-versus-time display

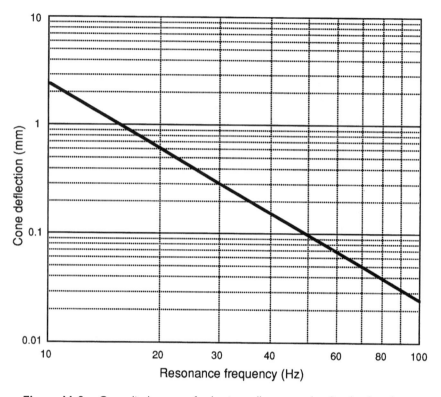

Figure 11-8. Cone displacement for horizontally mounted ceiling loudspeakers.

at any selected listening position. This shows the effects of reflections in the space and can be used to identify troublesome clusters of reflections.

Some programs also support an auralization program module that will enable the user to actually audition the target system (normally via binaural listening) while it is still on the drawing board.

Figure 11-9 shows an example of the coverage pattern of two 90° horns splayed horizontally. The gray scale at the right indicates the resulting levels at the listening plane. Figure 11-10 shows an example of level-versus-time at a selected listening position. Reflections here are calculated through the second order. Higher-order reflections may be examined as well, but the graphical plots showing sound rays tends to get very dense with succeeding orders of reflections.

11.6 System Equalization

Even though every effort is usually made to design a reinforcement system that is power flat over the bulk of its range, some degree of 1/3-octave band equaliza-

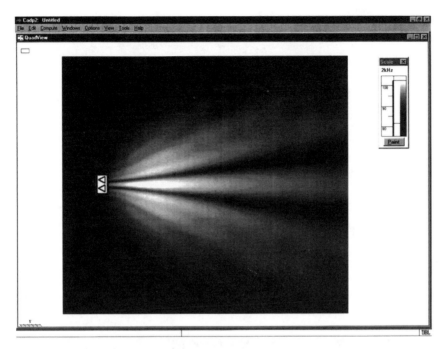

Figure 11-9. CADP2-estimated direct field coverage for two HF horns splayed in the horizontal plane. (Data courtesy JBL Professional.)

tion is normally applied to tailor the system to its acoustical environment, or perhaps to fine tune it to the tastes of the user. The measurement technique is similar to that discussed in Chapter 10, and the typical target equalization curve for sound reinforcement systems is shown as curve a in Figure 10-13.

Some consultants use a set of individually tuned narrow-band dip (notch) filters as a hedge against acoustical feedback at high system gain settings. Normally only a few of these initial feedback modes will be notched out. The curve shown in Figure 11-11*a* shows the response of both broad- and narrow-band equalization of a system. The resulting acoustical response of that system is shown in Figure 11-11*b*.

11.7 Measurements and Estimation of System Intelligibility

There are high expectations of sound reinforcement systems today, and there is a vast body of excellent engineering practice that supports the art. One of the most difficult remaining jobs, however, is that of estimating speech intelligibility performance that may be expected for a system that is still on the drawing board.

There are number of on-site measurement methods for determining the effectiveness of installed systems:

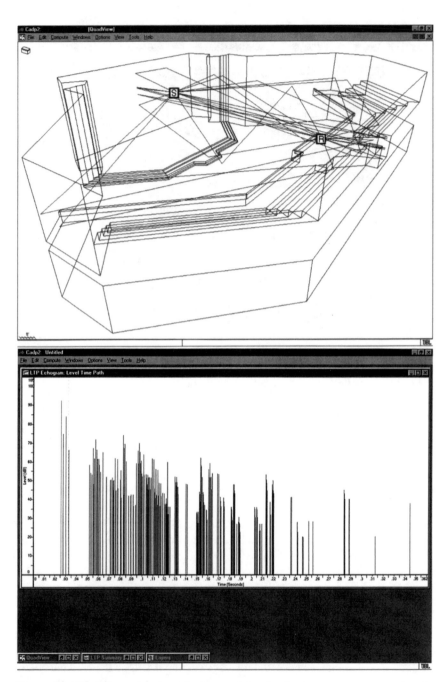

Figure 11-10. Ray tracing through the second order from a source to a receiver in an enslosed space (*a*); impulse response at a typical listening position due to first- and second-order reflections (*b*). (Data courtesy JBL Professional.)

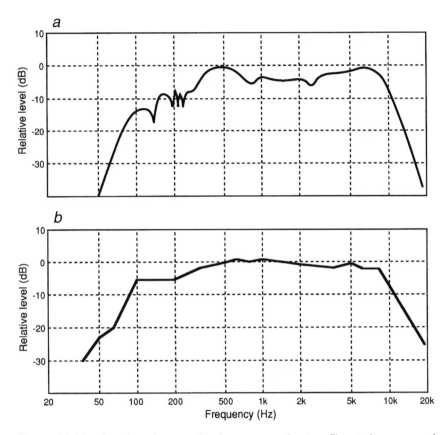

Figure 11-11. Broad- and narrow-band system equalization. Electrical response of filters (*a*); net acoustical response, measured on 1/3-octave intervals (*b*).

1. *Actual syllabic testing.* Subjects are asked to write down test words as they hear them in various parts of the hall. The test words are embedded in a carrier sentence so that the words to be identified are heard in the context of running speech. Many tests must be made with the same group of subjects in order to establish reasonable testing confidence limits.

In general, if a subject identifies 85% of the random test words, then that subject would be able to identify about 97% of all words in the sentence context of normal speech.

2. *Modulation Transfer Function (MTF).* Houtgast and Steeneken (1972) developed a method for measuring the effects of noise and reverberation on the integrity of speech. Their method of measuring these effects is to play a complex signal into the test space via loudspeaker and measure it at selected test points.

The signal is then analyzed, noting the reduction in modulation index of the signal caused by noise and reverberation effects. These components are weighted and a numerical rating is given to the system.

A simplification of this procedure, known as RASTI (*RA*pid *S*peech *T*ransmission *I*ndex) has been implemented into a convenient measuring system for on-site use. A small loudspeaker receives the test signal and is placed where the talker would ordinarily be. The receiving microphone is located at a target listening position and fed to the analyzer. The output is a numerical rating of the speech transmission index over that particular path.

3. *Clarity Index (Deutlichkeit)*. Thiele (1953) proposed a measurement of speech clarity based on early-late ratios of reflections of an impulse function delivered to a room. His basic equation is:

$$\text{Deutlichkeit} = \frac{\int_0^{50\,ms} [g(t)]^2 dt}{\int_0^{\infty} [g(t)]^2 dt} \times 100\% \tag{11.1}$$

where $g(t)$ represents the impulse response of the room.

Other workers in the field have devised methods for estimating intelligibility, without necessarily making measurements in the target space:

1. *Articulation Index (AI)*. Developed by French and Steinberg (1947) for use primarily in telephone circuits, the AI evaluates intelligibility in terms of signal-to-noise ratios in five weighted octave bands. The method, as modified by Kryter (1962), is shown in Figure 11-12.

2. *Percentage Articulation Loss of Consonants (%Alcons)*. Peutz (1971) determined that, in spaces where the speech-to-noise ratio was high, articulation loss of consonants could be determined by noting the reverberation time and direct-to-reverberant ratio in the octave band centered on 2 kHz. The basic relationships are shown in Figure 11-13.

The various procedures can be compared with each other, as shown in Figure 11-14. For example, the correlation between syllabic testing and AI is given in Figure 11-14*a*. The correlation between clarity index and measured speech intelligibility is shown in Figure 11-14*b*. While these figures show reasonable trends, the standard errors appear fairly large, often of the order of ±10%.

If it is possible, within reasonable limits, to estimate noise levels, reverberation time, and direct-to-reverberant ratios that will exist in yet unbuilt spaces, that data may be entered into the AI and %Alcons charts and a rough estimate of sound system intelligibility may be made. The charts may tell us that a system is going to be very good (or very bad), but confidence limits on marginal system performance may be rather wide.

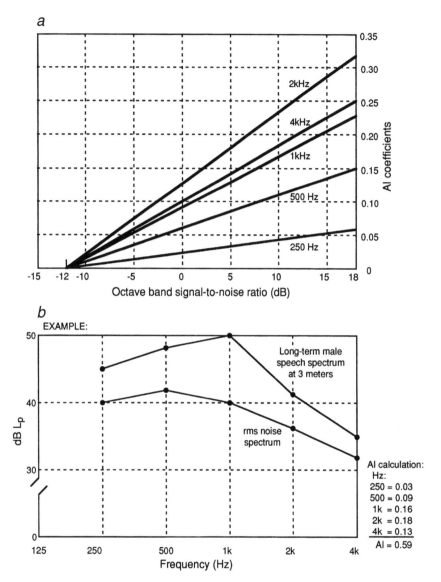

Figure 11-12. Calculation of AI. Determination of weighting factors (*a*); example of calculation (*b*).

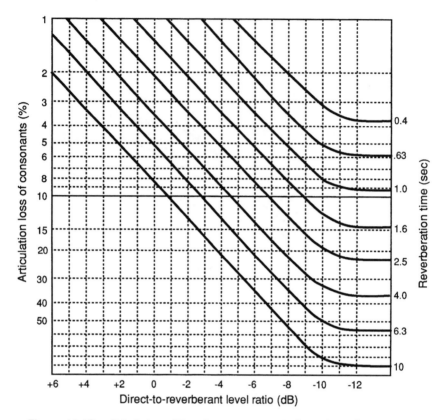

Figure 11-13. Calculation of Peutz's percentage articulation loss of consonants.

11.8 Electronic Halls

The term "electronic hall" refers to spaces whose acoustics have been enhanced via signal processing and loudspeakers. Many modern halls are multipurpose designs, and as such must provide well-damped acoustical properties for, say, lectures and motion pictures, as well as varying degrees of acoustical liveness for music performance. Electronic enhancement is one way of accomplishing this.

Other halls may be acoustically deficient in one regard or another, and electronic enhancement may represent a more attractive economical alternative to actual acoustical redesign of the hall.

11.8.1 Assisted Resonance

One of the earliest successful applications of electronics in architecture was the Assisted Resonance system installed in the Royal Festival Hall in London (Parkin, 1975). As originally designed, the hall lacked sufficient volume for its seating

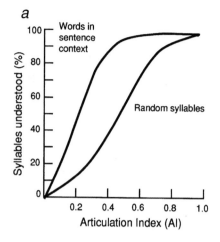

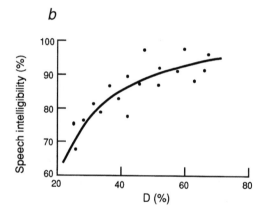

Figure 11-14. Correlation between intelligibility of random syllables and words in normal sentence context (*a*); correlation between clarity index and measured speech intelligibility (*b*).

capacity and target reverberation time. The Assisted Resonance system was installed to increase reverberation time in the frequency range from 60 to 700 Hz.

A total of 172 individual microphone-amplifier-loudspeaker channels were installed in the ceiling of the hall. The microphones were mounted in Helmholtz resonators so that each channel responded to a single narrow band of frequencies. Tunings were spaced so that they effectively overlapped in the desired frequency band. The microphones are located well away from direct sound sources and pick up primarily reverberant cues in the space.

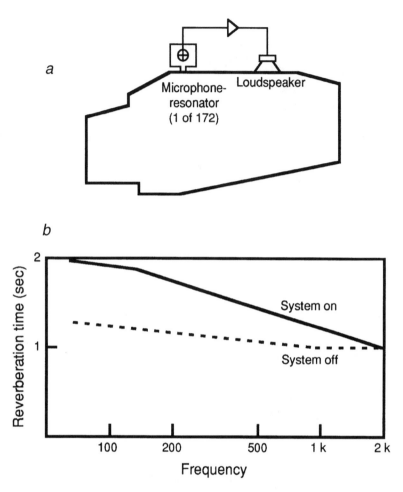

Figure II-15. Assisted resonance. System layout (*a*); reverberation time with and without system (*b*).

Because of their narrow-band response, the individual channels were stable, and the increase in apparent reverberation time resulted primarily from the high-Q tuning of each resonator. Details are shown in Figure 11-15.

11.8.2 Sound Field Amplification

Sound field amplification was a later development, and there are a number of directions that have been pursued. In the method shown in Figure 11-16, many individual wide-band microphone-amplifier-loudspeaker channels are arrayed about a performance space, more or less randomly, and each is operated at a

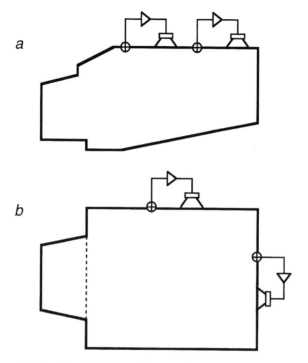

Figure 11-16. Sound field amplification. Section view (*a*); plan view (*b*).

relatively low gain setting (usually in the range of −15 to −20 dB). Properly adjusted, the ensemble of these amplification channels can produce the effect of a more reverberant target space—one of the same dimensions as the actual space in which the system is operating. In other words, the system effectively decreases the natural boundary absorption, resulting in an increase in reverberation time as well as a decrease in direct-to-reverberant ratio.

11.8.3 Lexicon Acoustical Reverberance Enhancement System (LARES)

LARES is a relatively recent development consisting of a number of microphone-reverberator-loudspeaker chains distributed about a space. For a large installation there would typically be two microphone inputs channels, 16 reverberators, eight amplifier channels, and upward of 40 loudspeakers, as shown in Figure 11-17. LARES is basically configured as a large distributed sound reinforcement system with the inclusion of digital reverberation generators, and as such can be operated for purposes of sound reinforcement as well as reverberation enhancement.

An important element in the performance of LARES is the use of randomly varying delay parameters in the operation of the reverberation elements, making

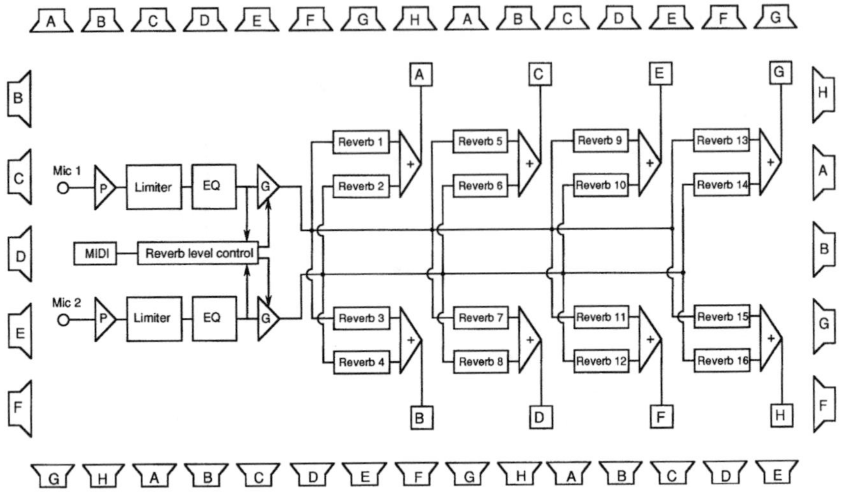

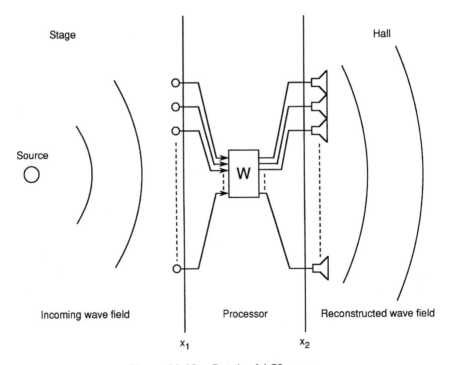

Figure 11-17. Details of LARES system.

Figure 11-18. Details of ACS system.

it difficult for acoustical feedback to occur at normal gain settings of the system. Since reverberation time can be set independently of system gain, LARES offers the user a degree of flexibility in meeting program demands that the other systems discussed here cannot match.

11.8.4 Acoustical Control System (ACS)

As described in Berkhout (1988), ACS, a horizontal line of microphones in front and over a performance stage, is connected with a line of house loudspeakers through a control matrix, as shown in Figure 11-18. This portion of the system amplifies the direct sound from the stage, and the subsequent wavefront reconstruction by the line of loudspeakers gives an accurate mapping of the stage sound sources for all listeners. An additional part of the system provides a number of microphone-delay-amplifier-loudspeaker loops located in the performance space to increase reverberation.

11.9 Environmental Effects on Sound Propagation

The propagation of sound can be profoundly influenced by wind, temperature, and humidity. These effects are aggravated all the more by the gradient effects that often exist out-of-doors.

11.9.1 Effects of Wind

The velocity of sound in air is the sum of its velocity in still air and the velocity of wind in a given direction. Moderate winds will have little effect, but pronounced wind velocity gradients, especially over large distances, affect the distribution of sound, as shown in Figure 11-19. The effect on the direction of sound in a strong cross breeze is shown in Figure 11-20.

11.9.2 Effects of Temperature

Sound travels faster in warm air than it does in cool air, and the gradient effects shown in Figure 11-21 are commonly observed in the early morning and early evening hours as the sun rises or sets. Over very large distances, sound may be observed to skip, creating quiet zones.

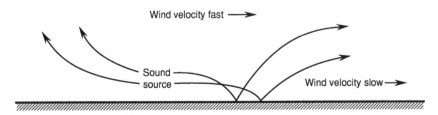

Figure 11-19. Effect of wind velocity gradient on sound propagation.

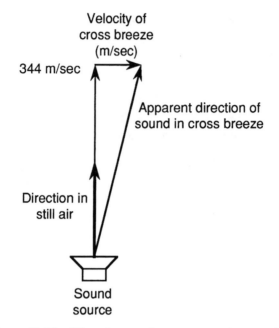

Figure 11-20.   Effect of a cross-breeze on sound propagation.

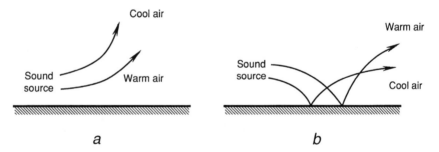

Figure 11-21.   Effect of temperature gradients on sound propagation.

11.9.3 Effects of Humidity

The effects of humidity on sound can easily be noticed indoors, and typical examples are shown in Figure 11.22. As seen in Figure 11-22a, the frequency spectrum of a sound reinforcement system will vary significantly over distance, especially if the relative humidity is fairly low.

The data shown in Figure 11-22b indicates the excess attenuation of sound with distance as a function of frequency and relative humidity.

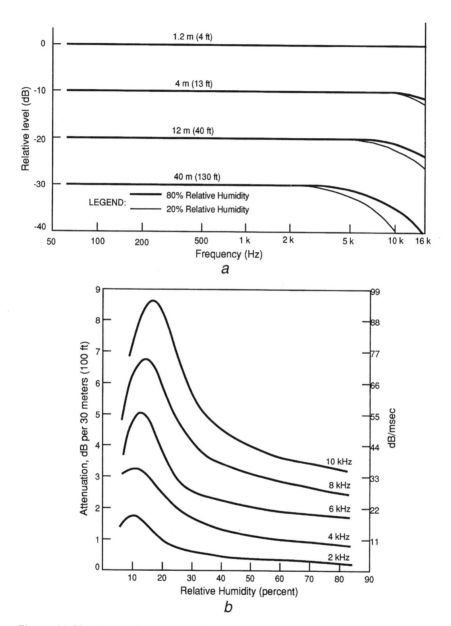

Figure 11-22. Atmospheric losses of sound: due to both inverse square relationships and humidity (*a*); relative attenuation of sound versus distance and of signal transit time as a function of relative humidity and frequency (*b*).

It is also worth noting that reverberation time will be shortened at high frequencies when the relative humidity is low.

11.10 System Stability

Throughout this chapter we have emphasized the importance of stability in sound reinforcement systems without defining the factors that govern it. *Stability* is defined here as the absence of electroacoustical feedback, a condition in which sound from the loudspeakers finds itself recirculating through the system microphones and back again through the loudspeakers. The following factors can minimize feedback:

1. Use of close operating distances between microphone and talker.
2. Locating microphones well outside the direct field coverage of loudspeakers.
3. Gating off any unused microphones.
4. Using well-behaved microphones—those whose pickup patterns are uniform over a wide frequency range and whose response is peak free.
5. Skillful employment of signal processing. For example, the LARES system discussed in Section 11.8.3 permits relatively high-reinforcement system gain through the use of random delay parameters that spoil the normal tendency of the system to go into electroacoustical feedback at high gain settings.

Bibliography

Berkhout, A., "A Holographic Approach to Acoustic Control," *J. Audio Engineering Society*, Vol. 36, No. 12 (1988).

Borwick, J., ed., *Loudspeaker and Headphone Handbook*, Butterworths, London (1988).

Davis, D., and Davis, C., *Sound System Engineering*, Sams, Indianapolis, IN (1987).

Eargle, J., *Electroacoustical Reference Data*, Van Nostrand Reinhold, New York (1994).

Eargle, J., *Handbook of Sound System Design*, Elar Publishing, Commack, NY (1989).

French, N., and Steinberg, J., "Factors Governing the Intelligibility of Speech Sounds," *J. Acoustical Society of America*, Vol. 19 (1947), pp. 90–119.

Houtgast, T., and Steeneken, H., "Envelope Spectrum and Intelligibility of Speech in Enclosures," presented at IEE-AFCRL Speech Conference, 1972.

Kryter, K., "Methods for the Calculation and Use of Articulation Index," *J. Acoustical Society of America*, Vol. 34 (1962), p. 1689.

Parkin, P., "Assisted Resonance," *Auditorium Acoustics*, Applied Science Publishers, London (1975), pp. 169–179.

Peutz, V., "Articulation Loss of Consonants as a Criterion for Speech Transmission in a Room," *J. Audio Engineering Society*, Vol. 19, No. 11 (1971).

Thiele, R., *Acustica*, Vol. 3 (1953), page 291.

Systems for Film and Video

12.1 Introduction

In a commercial sense, film and video entertainment dominate the world of audio, and yet are completely dependent on it. While motion picture sound has made steady strides over the decades, the relatively recent home theater phenomenon, driven in recent years by videotape and Laserdiscs, has literally redefined how most people spend their leisure hours at home and, by extension, how they listen to music.

For music-only presentation, the familiar stereo system, which has delivered music to us for nearly four decades, is slowly being replaced by a five-channel system, which is soon to become digital. This system is capable not only of delivering a convincing multichannel sound track for video, but also a new kind of music, known generally as *surround sound*. The loudspeakers have not changed significantly, but the playback environment has.

In this chapter we will examine the requirements for sound reproduction at the film production end, as well as observing how these techniques may be brought into the home with maximum effectiveness.

12.2 Motion Picture Loudspeaker Systems and Environment

12.2.1 Evolution of Loudspeakers

When sound was introduced into the motion picture in the late twenties, the requirement of filling a large room with adequate sound levels was hard to meet. Amplifiers of the day were usually in the 10-W range, and loudspeakers had to be as efficient as possible. No wonder that the early art depended almost entirely on the development of horn systems.

The first promising step occurred in the mid-thirties with the introduction of the Shearer-Lansing two-way system, shown earlier in Figure 7-19. The on-axis

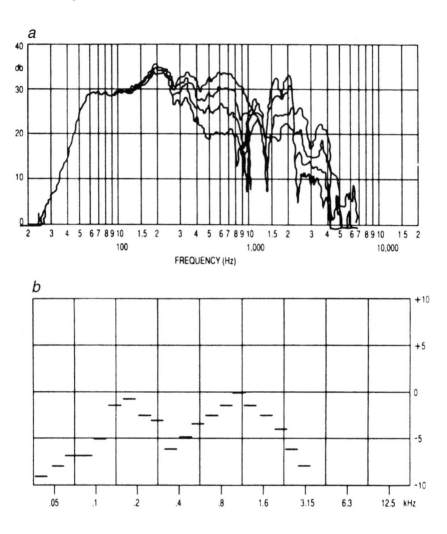

Figure 12-1. Response of LF ported horn systems. On- and off-axis response at 0°, 15°, 30°, and 45° (*a*); reverberant field response in the theater (*b*). [Data in (*a*) courtesy JBL, Inc.; (*b*) courtesy Dolby Laboratories.]

bandwidth of this system was fairly smooth from 50 Hz to 8 kHz, and the output capability was sufficient for the largest houses of the day.

During the forties, the LF portion of this design gave way to a hybrid ported horn system, similar to that shown in Figure 7-21*a* and *b*. The HF portion of the system remained as a multicellular horn-driver combination, crossing over from the LF section in the range of 500 Hz. While systems such as these could be coaxed into producing reasonable on-axis response, the power response of the systems was far from ideal, often exhibiting large "lumps" in response in the midbass and lower treble ranges.

Figure 12-1*a* shows a typical family of on- and off-axis response curves for these early ported LF horn systems. The reverberant response, or power response, of the complete system is shown in Figure 12-1*b*. The midbass peak in the 160–200 Hz range is clearly apparent in both sets of data, and the power response rolloff of the HF horn-driver combination can be seen as well.

Things did not change substantially until the mid-seventies when both the hybrid LF system and multicellular HF system gave way, respectively, to simple ported LF systems and uniform coverage HF horn systems. The improvement in on- and off-axis response provided by the new LF systems is shown in Figure 12-2*a* and *b*. The corresponding improvement in HF off-axis response can be seen by referring to Figure 7-14.

A very important aspect of these new power-flat systems is that the directional properties of both LF and HF sections are nearly the same at the crossover frequency of 500 Hz. The horizontal beamwidth is nominally 90°, providing wide coverage in the theater, and the vertical beamwidth is nominally 40°, providing tight coverage over the entire audience area, as seen from the screen. Such systems can be equalized for a desired overall response, and that response will be relatively uniform throughout the house. The importance of narrow vertical pattern control is to minimize the amount of sound that impinges directly on the walls and ceiling of the theater, which would generate needless reverberant effects.

12.2.2 Evolution of Theater Architecture and Acoustics

Early motion picture houses were often converted vaudeville houses, and as such had deep stages, large orchestra pits, and steep balconies. These large rooms often seated 1000–1500 or more people and were acoustically fairly live in order to support unamplified music and speech. From the viewpoint of acoustics for sound motion pictures, there was much to be desired.

As the art of the motion picture evolved after World War II, dedicated houses were built for film. In time, the balcony vanished, and capacity was generally restricted to no more than 800–1000 patrons. With the coming of Dolby noise reduction in film sound tracks, attention was given not only to loudspeakers but also to the control of acoustics in the house itself. Better structural isolation and use of architectural damping made for quieter spaces, and the sound tracks could thus convey a wider range of music, dialog, and effects for the sake of the picture itself.

The modern motion picture theater is usually designed for reverberation times in the range shown in Figure 12-3. The average house volume is about 5 m^3 (190 ft^3) per patron, so it can be seen that a house seating 1000 patrons will

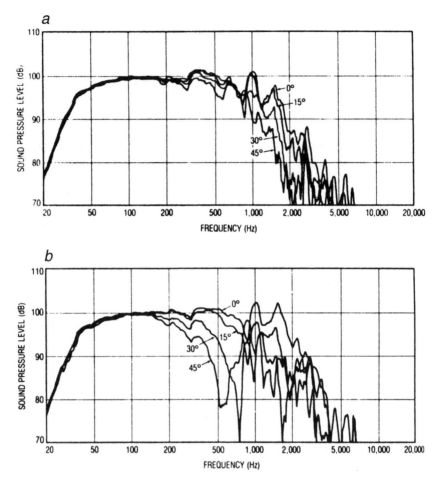

Figure 12-2. On- and off-axis response of dual-driver ported LF system: horizontal (a); vertical (b) (Data courtesy JBL, Inc.).

have a relatively short reverberation time, certainly as compared with a music performance space of the same seating capacity. The back wall of the theater is normally made very absorptive in order to reduce the level of reflections back toward the screen.

Care should be taken in the theater architectural design to avoid discrete reflections, especially from the side walls, that would tend to interfere with dialog intelligibility. In general, the pattern of reflections in the theater should be neutral, not in any way suggesting a sense of large space. If called for, such effects are picture dependent and would be conveyed through the surround loudspeaker systems in the house.

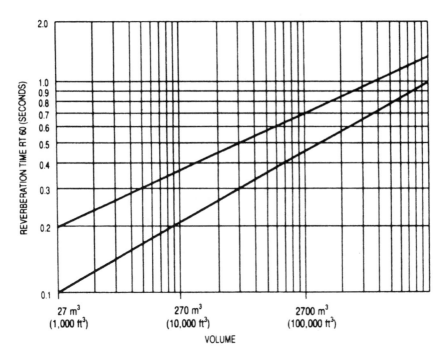

Figure 12-3. Optimum reverberation time in motion picture theaters as a function of room volume.

Figure 12-4. A large theater installation. (Photo courtesy JBL. Oscar© statuette A.M.P.A.S.®.)

12.3 System Layout in the Theater

12.3.1 Electrical Considerations

Figure 12-4 shows a photograph of the main loudspeakers in a large house seating about 1000 persons before installation of the perforated screen. Note that the

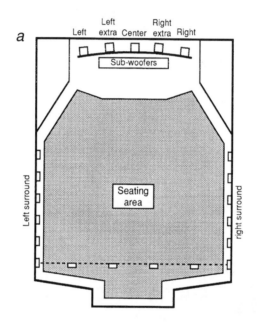

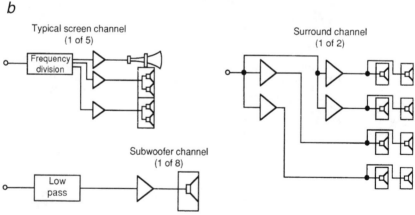

Figure 12-5. A large theatre installation. Room plan view (*a*); electrical diagram (*b*).

loudspeakers are mounted in a large wall that extends from floor to ceiling and from side to side in the space. The five screen channels shown here are primarily for exhibiting the 70-mm six-track magnetic format, which is not being used for new releases at the present time. Today the normal format used in the theater consists of three screen channels, two surround channels, and a LF channel for special effects.

A typical loudspeaker layout for a large theater is shown in plan view in Figure 12-5*a*, with the corresponding electrical flow diagram shown in Figure 12-5*b*. Note that the five screen loudspeaker arrays are triamplified. In normal practice, the lower LF sections roll off above 315 Hz, while the upper LF sections cross over into the HF horns at 500 Hz. Both LF sections continue downward in response to about 40 Hz. (Note: in many new theaters, only three screen channels—left, center, and right—are installed.)

The very-low-frequency (VLF) requirements are taken care of by a group of subwoofers, each driven by a separate amplifier from a common LF program source. This channel is used for the normal LF extension of program as well as for special effects, with response that can be maintained flat down to 25 Hz.

The screen systems are equalized using a multiple microphone pickup technique in the house. The standard curve to which each loudspeaker is equalized is as shown in Figure 12-6; it is standardized by the International Standards Organization (ISO) and is known as the *x-curve*. Equalization takes place through the screen, thus taking into account the HF screen losses. The degree of loss depends on screen material, percentage perforation, and the specific angle of transmission through the screen. Typical on-axis loss for 8% perforation through a 0.3-mm thick screen is shown in Figure 12-7.

It is customary practice today to equalize the screen systems out to 16 kHz in accordance with the standard curve, and this requires considerable boosting of the HF signal to the compression drivers, often on the order of 10 dB at the highest frequencies.

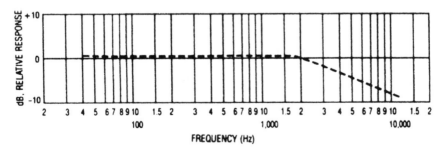

Figure 12-6. ISO Bulletin 2969–recommended response curve for motion picture loudspeaker systems,

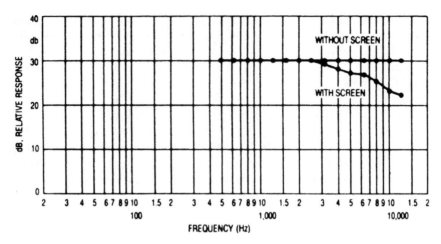

Figure 12-7. Typical screen losses (see text for details).

12.3.2 Standard Playback Levels in the Theater

Throughout the professional motion picture industry, both at the studio and public exhibition stages of the process, playback levels are fairly well controlled. Considering digital sound tracks, each screen channel is set so that a nominal digital signal level of −20 dBFS (dB relative to full scale) will produce an average level in the house of 85 dB Lp. Dialog is normally reproduced at peak levels of 85 dB, while the remainder of the 20 dB of headroom is used for music and sound effects. Each screen channel is specified to handle broad-band signal levels at 0 dBFS, thus producing peaks in the range of 105 dB. An ensemble of three screen channels can thus produce a level about 5 dB higher. Through careful aiming of the HF horns, HF level variations in the seating area from front to back in the theater can be kept within a range of ±3.5 dB.

The subwoofer channel is normally capable of greater sound pressure levels, due primarily to the equal loudness contours at low frequencies, as shown in Figure 12-8. As an example, for a 1 kHz level of 85 dB Lp, a signal at a frequency of 25 Hz will have to be about 112 dB in order to sound as loud as the 1 kHz signal. At a midband reference level of 105 dB, the 25 Hz signal will have to be presented at a level of about 125 dB for equal loudness!

Clearly, such levels are very difficult to generate in a large theater, but the situation does point out the necessity of some degree of elevated peak level capability for the subchannel relative to the screen channels. In general, if the subwoofer channel can deliver levels of 115 dB at 25 Hz it is considered excellent indeed.

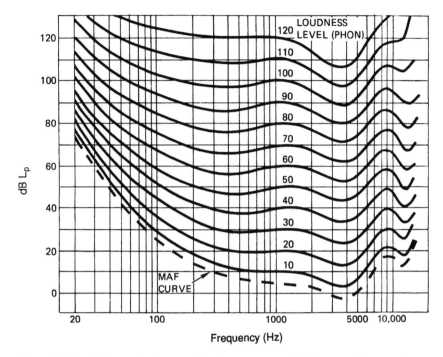

Figure 12-8. Robinson-Dadson equal loudness contours. MAF curve indicates *minimum audible field.*

12.3.3 Audio Formats in the Motion Picture Theater

Numerous formats have been used in motion pictures over the years, including monophonic optical and matrixed stereo optical tracks, as well as four- and six-channel magnetic tracks. Since the early nineties, digital formats have become the norm, providing up to five screen channels (three is most common), along with two surround sound channels and a special effects channel.

From the point of view of loudspeaker design and choice, the digital formats have brought very stringent requirements, inasmuch as these formats all exhibit flat power bandwidth at the highest frequencies. The electrical capability of various Dolby motion picture recording systems can clearly be seen in Figure 12-9.

12.4 Specific Loudspeaker Models for the Motion Picture Theater

Figure 12-10 shows a group of loudspeaker products intended for general motion picture use. In this age of multiplex theaters, there are many small theaters, often

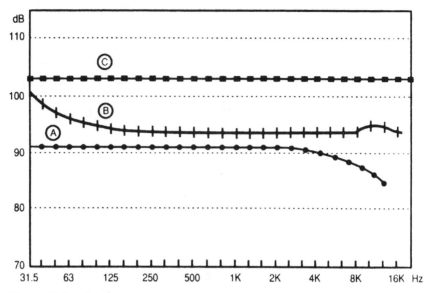

Figure 12-9. Relative electrical headroom in Dolby motion picture recording systems. Curve A: Dolby A-type noise reduction; curve B: Dolby-SR spectral recording; curve C: Dolby AC-3 digital recording. (Data courtesy Dolby Laboratories.)

Figure 12-10. Photograph of various loudspeakers intended for motion picture use. (Data courtesy JBL, Inc.)

seating as few as 50 patrons. In these spaces, loudspeaker requirements are reduced and many small systems are available to fill these needs. The loudspeaker systems normally used for surround channel application in smaller and midsize houses are essentially modifications of three-way home high-fidelity models. Surround loudspeakers used in large houses are normally small systems with horn HF sections for greater output capability. In general, theater loudspeaker layout should ensure that each of the two surround channels are capable of midband acoustical output capability that is a good match for the midband output capability of a single screen channel. This requirement often dictates that upward of 8–12 loudspeakers are used in each surround channel in order to provide the desired power matching with the main loudspeakers.

12.5 Multichannel Video in the Home

The home theater revolution of the early nineties was based largely on the wide availability of Dolby Stereo encoded analog tracks on both Laserdiscs and VHS videotapes of current motion pictures. At first, a handful of hardware manufacturers began making audio-video (AV) receivers, which provided five channels of amplification along with front-end processing using an integrated circuit Dolby matrix decoder. Such an arrangement produced left, center, right, and surround outputs derived from the basic stereo signal.

The Dolby stereo matrix had earlier made its entry into the motion picture world as Dolby Stereo Optical, an economical alternative to magnetic tracks on 35- and 70-mm film. This same stereo pair of magnetic tracks subsequently drove the homes market for surround sound video.

The translation of the theater experience into the home began in a casual way. Manufacturers at first provided small loudspeaker models for the purpose so as not to upset traditional notions of home decor, but these loudspeakers did not necessarily do justice to modern action films.

12.5.1 Frontal Loudspeakers

Much of the early guidance for consumer manufacturers in this area was provided as a licensing service by the THX division of Lucasfilm, a major producer of films. THX had earlier promulgated audio and optical performance standards for commercial motion picture theaters and was now utilizing that in the home theater market. From the loudspeaker point of view, THX proposed that the accurate translation of the motion picture experience into the home called for frontal loudspeakers with the same wide horizontal pattern control and relatively narrow vertical midrange control characteristic of motion picture theater systems. In the theater, the directional control is normally provided by a horn, which can control horizontal and vertical radiation patterns independently of each other over a large frequency range. By comparison, consumer loudspeaker systems are generally composed of cone and dome drivers, and as such are subject to the diameter-dependent radiation patterns of those devices as described in an earlier chapter.

In order to meet the narrow vertical coverage requirement, THX specifies MF transducers as vertically arrayed pairs. A frontal view of a typical wide range frontal loudspeaker array is shown in Figure 12-11a. Note that the center loudspeaker, since it must be placed above or below a television screen, has a relatively narrow vertical profile.

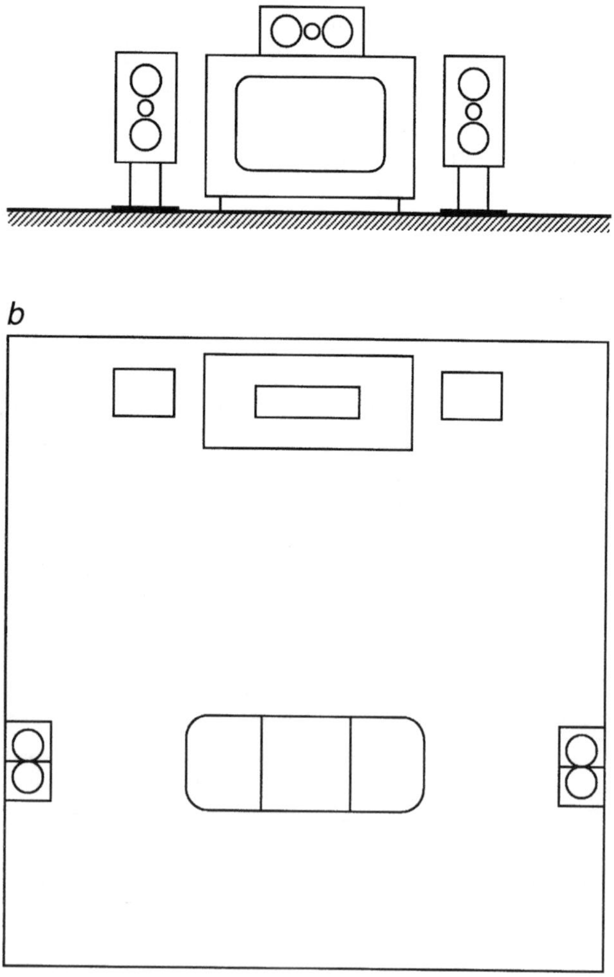

Figure 12-11. Typical baffle layout for front channels of a home theater system (a); plan view of the home theater setup showing dipole surround loudspeakers (b).

12.5.2 Surround Loudspeakers

Surround channel requirements are more difficult to define. In the theater, a large number of surround loudspeakers are used; for a typical listener the surround impression is the result of many loudspeakers, whose individual sounds arrive at the listener's ears with different levels, directions, and delay times. This arrangement results in a degree of sound diffusion having a timbre and general spatial impression that may be difficult to duplicate in the typical living room. In its attempt to match the theater sound experience, THX uses single-dipole loudspeakers on each side of the listening room with their null axes aimed at the primary listening position, as shown in Figure 12-11*b*. In this arrangement, the major contribution of the surround channels comes at the listener by way of first-order room reflections, not directly from the surround loudspeakers themselves. While this may not be an exact translation of the theater experience into the home listening environment, it does provide a good approximation. For direct ambient music reproduction, the dipole arrangement provides an adequate timbral match with the front channels if the listening space is acoustically well damped acoustically.

12.5.3 Low-Frequency Requirements

The attractiveness of subwoofers in the home is that a single subwoofer can often deliver the LF output necessary to complement a five-channel full-range loudspeaker array. Thus, the size requirement of the individual five-channel array can be substantially reduced, since the directionality of frequencies below about 100 Hz is difficult to assess. Subwoofers are most easily integrated into home

Figure 12-12. Photograph of typical home video loudspeaker. (Courtesy Citation.)

environments if they are self-powered and fed from a dedicated output on the basic AV receiver.

12.5.4 General Comments

Not all manufacturers are in agreement that home theater loudspeakers should necessarily emulate the character of theater loudspeaker systems. Strongly dissenting voices state that what has traditionally been good for music should be good for home video as well, so there is ample room for individual taste. Figure 12-12 shows a typical loudspeaker array intended for home video reproduction.

Bibliography

Allen, I., *Technical Guidelines for Dolby Stereo Theatres*, Dolby Laboratories, San Francisco (1993).

Eargle, J., *Electroacoustical Reference Data*, Van Nostrand Reinhold, New York (1994).

Eargle, J., *Handbook of Recording Engineering*, Chapman & Hall, New York (1996).

Eargle, J., Bonner, J., and Ross, D., "The Academy's New State-of-the Art Loudspeaker System," *J. Society of Motion Picture and Television Engineers*, Vol. 94, No. 6 (1985).

Eargle, J., and Means, R., "A Microcomputer Program for Determining Loudspeaker Coverage in Motion Picture Theaters." *J. Society of Motion Picture and Television Engineers*, Vol. 93, No. 8 (1984).

Engebretson, M., and Eargle, J., "Cinema Sound Reproduction Systems," *J. Society of Motion Picture and Television Engineers*, Vol. 91, No. 11 (1982).

International Organization for Standardization (ISO) Bulletin No. 2969.

Cinema Sound System Manual, JBL, Inc., Northridge, CA (1995),

Motion Picture Sound Engineering, D. Van Nostrand Company, New York (1938).

Loudspeaker Measurements and Modeling

CHAPTER

13

13.1 Introduction

Loudspeaker measurement technology has grown significantly in the last two decades since digital signal processing has come on the scene. In earlier days, we had mechanically driven sine wave oscillators operating in synchronism with moving paper chart recorders, many of which are just now winding down their useful existence. With specific regard to loudspeakers, we are still learning just what performance attributes need to be measured, and to what degree of detail.

At the transducer design stage, there are matters of basic mechanical system linearity and integrity, which are often preceded by detailed modeling by the design engineer. Here, it may be helpful to examine cone and diaphragm movements via stroboscopic and laser methods so that higher-order performance anomalies can be analyzed. Later, when drivers are assembled into systems, acoustical measurements will dominate the design process.

We will begin with a discussion of analog frequency-based measurements.

13.2 Frequency Response Measurements

Figure 13-1 shows details of the most common method of running frequency response curves. A beat frequency oscillator (BFO) is motor driven and puts out a constant level sine wave signal that normally covers the frequency range from 20 Hz to 20 kHz. The sweep is logarithmic, so that that the sweep time is constant per octave or other bandwidth percentage. The BFO signal is fed to the device under test, and a measurement microphone is placed at a fixed reference distance. The signal from the microphone is fed to a graphic level recorder where it is converted to a level in dB and plotted via a vertically moving pen on a horizontally

257

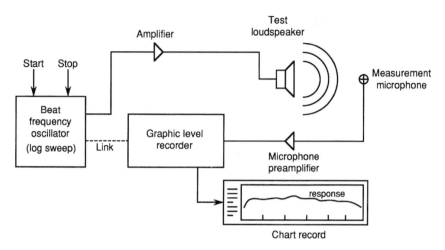

Figure 13-1. Block diagram for analog measurement of frequency response using a beat frequency oscillator.

moving strip of paper. The paper movement is in exact synchronism with the signal generator through either electrical or mechanical interlocking.

The important variables in operating the system are setting the paper speed and the pen writing speed to ensure good data. Examples are shown in Figure 13-2a and b. In Figure 13-2a, the paper speed is slow and the writing speed fast, producing the greatest detail; in Figure 13-2b, both speed and the writing speed are slow, producing highly averaged data. Most measurement systems have a range of 40–50 dB. The data in Figure 13-2a may be valuable to the transducer designer, while the data in Figure 13-2b may be more representative of subjective effects as perceived by the listener.

The system can be varied slightly to produce averaged data. Figure 13-3 shows a configuration for driving a tracking bandpass filter, with a constant pink noise signal fed to the loudspeaker. The method can be varied further so that 1/3-octave noise bands are delivered sequentially to the loudspeaker under test.

The magnitude (or modulus) of a loudspeaker's impedance can be plotted by measuring the voltage drop across the loudspeaker when the loudspeaker is driven with a constant current source. In the circuit shown in Figure 13-4, the system is calibrated by inserting a 10-Ω resistor and adjusting the gain of the graphic level recorder so that its pen indicates the 10-Ω marker on the chart paper. When the proper gain has been set, the calibration resistor is switched out and the loudspeaker load inserted in its place. The impedance plot generated in this manner is a plot of log impedance, rather than a linear representation. This convention is used throughout the book.

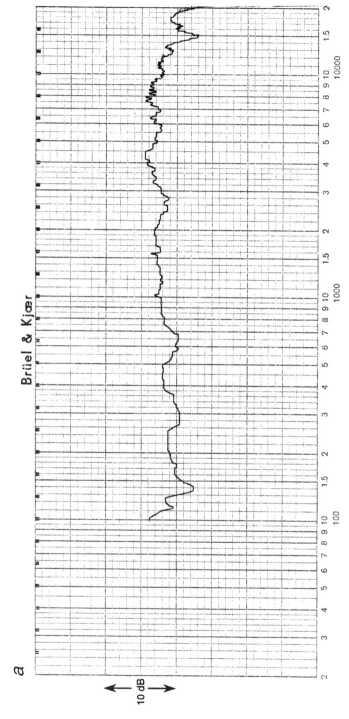

Figure 13-2. Effect of writing and paper speed on data resolution, shown on Brüel & Kjær standard chart recorder. Fast writing speed and slow paper speed (*a*); slow writing speed and slow paper speed (*b*). (Data courtesy JBL and Gregory Timbers.)

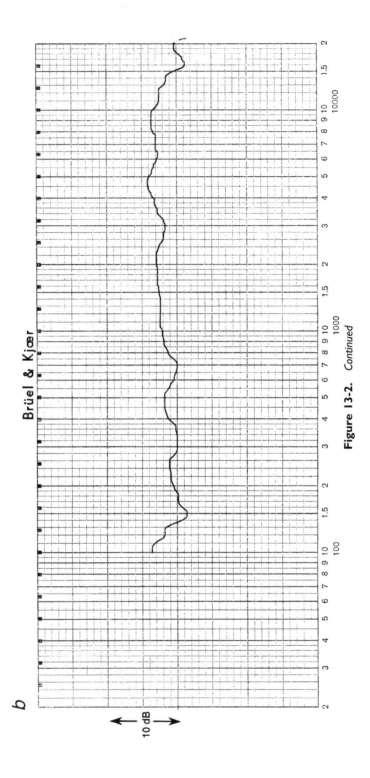

Brüel & Kjær

10 dB

b

Figure 13-2. *Continued*

260

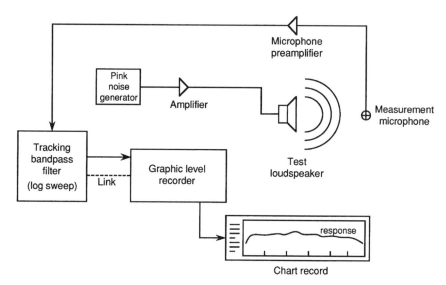

Figure 13-3. Block diagram for analog measurement of frequency response using a pink noise source and variable-bandpass analyzer.

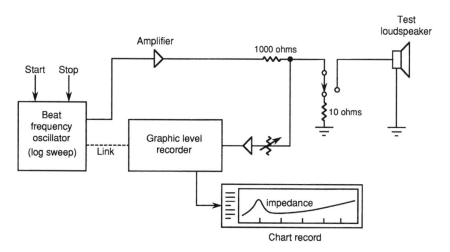

Figure 13-4. Measurement of loudspeaker modulus of impedance.

13.3 Distortion Measurements

There are a number of relevant distortion measurements that may be made on loudspeakers. The most common is to drive the loudspeaker with a swept sine wave at various nominal power inputs and measure the relative amounts of second

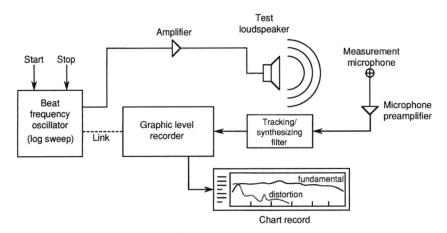

Figure 13-5. Measurement of harmonic distortion.

and third harmonic components, relative to the fundamental component. The setup for this is shown in Figure 13-5.

In normal practice, a frequency run of the fundamental is first made. Then the chart paper is rewound and a tracking-synthesized filter is used to generate the second harmonic passband; that signal is then plotted. The paper is once again rewound and the third harmonic plotted. In normal practice, the gain of the recorder is increased 20 dB when plotting the distortion components, so that they may be clearly seen on the rather narrow (40 or 50 dB) vertical window of the display.

When using a microphone with an upper frequency limit of 20 kHz, it should be clear that second-harmonic components cannot be read for fundamentals higher than 10 kHz, or third-harmonic components higher than about 6700 Hz.

Another way of looking at distortion is the THD + N (total harmonic distortion plus noise) method. Here, a tracking band rejection filter is used to notch out the fundamental frequency, leaving only the power summation of harmonics and of course any noise that may be present. This method is more applicable to electronic systems than to loudspeakers.

Most harmonic distortion in loudspeakers is related to mechanical nonlinearities, and this produces amplitude distortion. In some systems, where cone displacement is excessive, there may be frequency modulation (FM) effects as the cone's velocity becomes a significant fraction of the speed of sound. The measurement methods discussed here do not distinguish between the two, and a frequency modulation discriminator is necessary to directly isolate the FM components. In general, amplitude modulation effects predominate in normal loudspeaker performance.

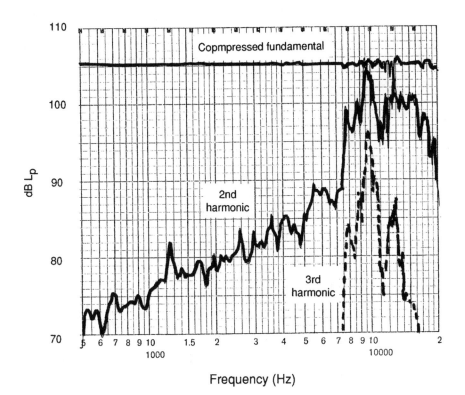

Frequency (Hz)

Figure 13-6. Illustration of the use of the compressed fundamental in distortion measurements. Distortion is raised 20dB. (Data courtesy JBL and David Bie.)

13.3.1 Use of Compressed Fundamental

Some measurement systems have a compressor circuit that can be used keep the fundamental level flat at the measurement microphone. This feature is of considerable use when comparing different transducers under identical output conditions. An example is shown in Figure 13-6, where the fundamental level for a compression driver-horn combination has been maintained at 105 dB at a distance of 1 m. We can clearly see the steady increase in second-harmonic distortion at a rate of 6 dB per octave. Above about 7 kHz, the distortion rises precipitously.

13.3.2 Intermodulation Distortion Measurements

Intermodulation (IM) distortion results from a combination of two sine waves, and the sum and difference combination tones that are generated by system

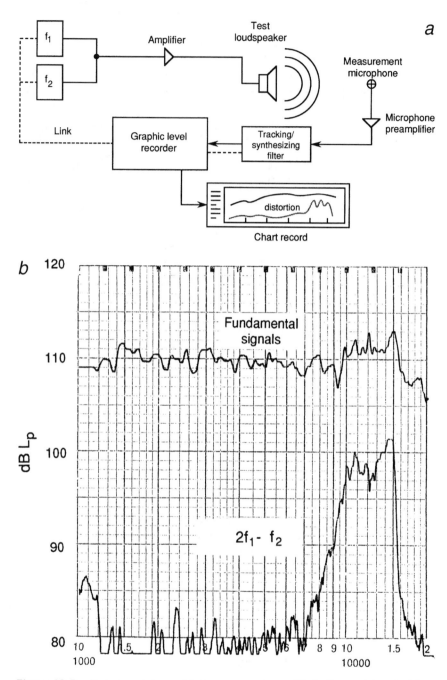

Figure 13-7. Twin-tone intermodulation measurements. Block diagram (*a*); typical data presentation (*b*). Distortion is raised 20 dB.

264

nonlinearities are plotted. Many of the various combination tones can be analyzed by available tracking filter-analyzer systems. Figure 13-7*a* shows the basic measurement setup.

The attractiveness of IM distortion measurement is that it relates, at least in part, to the way systems behave on actual program material. As such it may be a good indicator of when problems will occur, but it may not identify the specific cause of the problem as easily as single frequency measurements will. Today, IM distortion is normally carried out by establishing a fixed frequency difference between two sine waves, and then sweeping the pair of tones over the entire frequency band. If f_1 is one of the tones and f_2 the other, then $f_1 - f_2$ will be a constant; it is normally set in the range from 200 Hz to 1 kHz.

Due to nonlinearities in the system, several combinations of tones will be generated; the most significant are: $f_1 - f_2$ and the second-order tones $2f_1 - f_2$ and $2f_2 - f_1$. Figure 13-7*b* shows the output of the $2f_2 - f_1$ difference tone under the same drive conditions as the data in Figure 13-6. In the data shown here, there is very little distortion until just above 7 kHz, where it rises quickly.

13.4 Phase and Group Delay Response of Loudspeakers

Traditionally, phase and group delay response of loudspeakers have been difficult measurements to make, but the newer transform-based digital systems have made the process much easier. The older method of making phase measurements is shown in Figure 13-8. Because of the acoustical delay path between the loudspeaker and microphone, the oscillator output to the phase meter must be delayed by the same amount of delay. If this is not done, the phase meter will register multiple-phase rotations, especially at high frequencies, and the data will be very

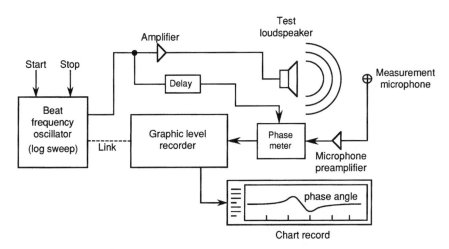

Figure 13-8. Analog measurement of phase response.

difficult to interpret. With digital delay this adjustment can be made easily and good data taken. The reader can readily understand that such measurements as these were all but impossible in the predigital era.

Group delay is the derivative of the phase response with respect to frequency, $-d\phi/d\omega$, and in the early days was usually calculated by inspection and graphical differentiation of the phase plot. Today's instrumentation presents delay data directly.

13.5 Measurement of Directional Data

Although we are primarily interested in the directional response of a loudspeaker system over its nominal frontal radiation angle, several applications of system modeling require a complete spherical description of the loudspeaker's directional response—and at many different frequencies. The traditional way of showing directional data is by way of the polar graph. The Brüel & Kjær standard chart recorder, long the mainstay of the electroacoustics industry, can accommodate a circular piece of sprocketed graph paper, driving it 360° in synchronism with a turntable that carries the loudspeaker. The basic setup is shown in Figure 13-9a. A typical polar graph of a loudspeaker is shown in Figure 13-9b.

There are a number of derived methods for presenting directional data for a loudspeaker, as shown in Figure 13-10.

1. *Frontal isobars.* This method gives a quick picture of the frontal characteristics of a device and is very handy in the specification of sound reinforcement components. A separate isobar plot is necessary for each frequency band.

2. *Family of off-axis curves.* These are often presented with the on-axis curve normalized, as shown here, and are generally given only in horizontal and vertical planes.

3. *Horizontal and vertical beamwidth (−6 dB) plots.* These methods are useful in sound reinforcement applications where they clearly indicate to the specifying engineer regions of loss of pattern control or excessive narrowing of pattern control.

4. *Plots of directivity index and directivity factor.* These plots give a single value at each frequency that enable certain rapid calculations to be made in many aspects of sound reinforcement.

5. *Spherical data file.* A complete data file for an arbitrary device will consist of 180° of front-to-back information for the entire 360° of rotation about the frontal axis. The data is normally measured at 5° or 10° increments and is normally expanded out in tabular form on spherical coordinates. A separate file is needed for each frequency, or frequency band, of interest. In many cases, data packing reduces the space actually occupied by one of these files. For example, device symmetry about one or two axes can reduce the requirements.

The amassing of this degree of data used to be a laborious process. Today,

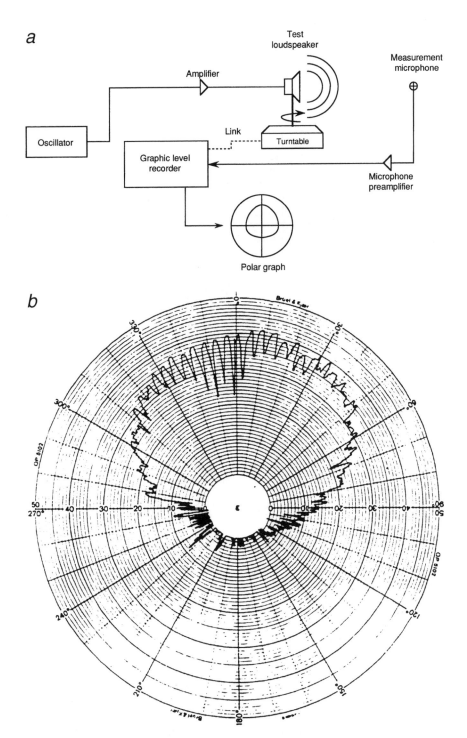

Figure 13-9. Polar graphs. Block diagram (*a*); typical data presentation (*b*).

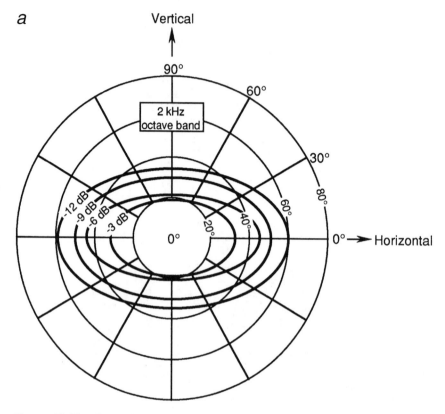

Figure 13-10. Derived methods of directional data presentation. Frontal isobars (*a*); off-axis response (*b*); beamwidth (−6 dB) versus frequency (*c*); directivity index (DI) (*d*); spherical data file (*e*).

there are computerized methods of mechanically indexing large devices, along with multimicrophone arrays, for gathering much data through parallel processing. A typical tabular file is shown in Figure 13-10. Here, ϕ (phi) represents the off-axis rotation from front to back of the device, while θ (theta) represents rotation about the frontal axis of the device from horizontal to vertical.

13.6. The Measuring Environment

The traditional measurement environments are anechoic chambers, open fields, and reverberant rooms. A section view of an anechoic chamber is shown in Figure 13-11. The wedges are made of medium density fiberglass and form a gradual lossy path from the interior of the chamber to the outer wall of the chamber. The wedges should be about 1/4 wavelength deep at the lowest frequency the chamber is intended to operate at. Furthermore, there must be adequate

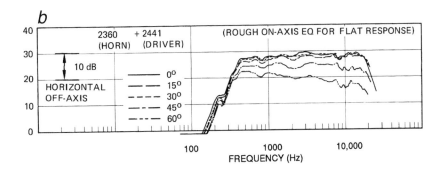

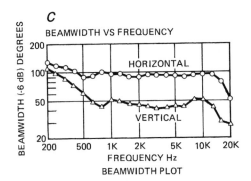

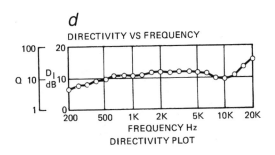

Figure 13-10. *Continued*

working space in the chamber so that LF measurements can take place at least 1/4-wavelength away from the wedge tips.

Anechoic chambers are very expensive, and many tens of thousands of dollars may be involved even in a modest one. A chamber that works well down to 100 Hz is adequate, since performance below that frequency can be accurately estimated.

The open field, often called a *ground plane*, is shown in Figure 13-12. It is

```
LOUDSPEAKER FILE:    jbl2360
FREQUENCY BAND:      0-5 kHz
DIRECTIVITY INDEX:   9
SENSITIVITY:         112
MAXIMUM POWER:       50
```

e

UPPER HEMISPHERE DB(THETA,PHI)

PHI / THETA	0o	10o	20o	30o	40o	50o	60o	70o	80o	90o
0o	0	0	-2	-3	-5	-6	-8	-10	-13	-14
10o	0	0	-1	-2	-5	-6	-8	-11	-13	-15
20o	0	0	-2	-3	-5	-7	-10	-12	-14	-17
30o	0	-1	-2	-3	-6	-9	-11	-13	-17	-19
40o	0	-1	-2	-4	-6	-9	-12	-14	-17	-20
50o	0	-1	-2	-4	-7	-10	-12	-15	-18	-21
60o	0	-1	-2	-4	-8	-12	-13	-14	-18	-20
70o	0	-1	-2	-5	-8	-11	-13	-15	-17	-19
80o	0	-1	-2	-5	-9	-11	-13	-15	-17	-20
90o	0	-1	-3	-5	-9	-12	-13	-15	-17	-19
100o	0	-1	-2	-5	-9	-11	-13	-15	-17	-20
110o	0	-1	-2	-5	-8	-11	-13	-15	-17	-19
120o	0	-1	-2	-4	-8	-12	-13	-14	-18	-20
130o	0	-1	-2	-4	-7	-10	-12	-15	-18	-21
140o	0	-1	-2	-4	-6	-9	-12	-14	-17	-20
150o	0	-1	-2	-3	-6	-9	-11	-13	-17	-19
160o	0	0	-2	-3	-5	-7	-10	-12	-14	-17
170o	0	0	-1	-2	-5	-6	-8	-11	-13	-15
180o	0	0	-2	-3	-5	-6	-8	-10	-13	-14
190o	0	0	-1	-2	-5	-6	-8	-11	-13	-15
200o	0	0	-2	-3	-5	-7	-10	-12	-14	-17
210o	0	-1	-2	-3	-6	-9	-11	-13	-17	-19
220o	0	-1	-2	-4	-6	-9	-12	-14	-17	-20
230o	0	-1	-2	-4	-7	-10	-12	-15	-18	-21
240o	0	-1	-2	-4	-8	-12	-13	-14	-18	-20
250o	0	-1	-2	-5	-8	-11	-13	-15	-17	-19
260o	0	-1	-2	-5	-9	-11	-13	-15	-17	-20
270o	0	-1	-3	-5	-9	-12	-13	-15	-17	-19
280o	0	-1	-2	-5	-9	-11	-13	-15	-17	-20
290o	0	-1	-2	-5	-8	-11	-13	-15	-17	-19
300o	0	-1	-2	-4	-8	-12	-13	-14	-18	-20
310o	0	-1	-2	-4	-7	-10	-12	-15	-18	-21
320o	0	-1	-2	-4	-6	-9	-12	-14	-17	-20
330o	0	-1	-2	-3	-6	-9	-11	-13	-17	-19
340o	0	0	-2	-3	-5	-7	-10	-12	-14	-17
350o	0	0	-1	-2	-5	-6	-8	-11	-13	-15

Figure 13-10. *Continued*

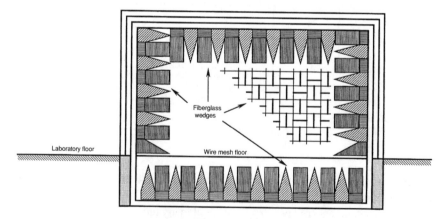

Figure 13-11. Section view of an anechoic chamber.

Open air

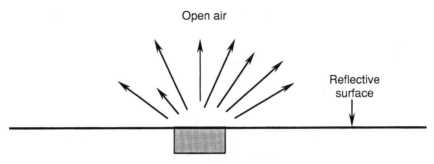

Reflective
surface

Figure 13-12. Section view of an open field.

often built on a rooftop, where there are few space limitations. Surfaces 10 m on a side are common and permit excellent LF measurements to be made. On the debit side, there are weather and noise to contend with. Tracking filters can help under noisy conditions, but bad weather puts these facilities out of business.

Outside of large research facilities, reverberant chambers are not found these days. They may have outlived their usefulness in loudspeaker measurement, given the many improved methods of data gathering that are common today. The virtue of the reverberant space is that it provides a real-time spatial integration of the signal, facilitating power response and efficiency measurements.

Today the listening room itself has become a measuring environment. Many modern techniques provide for signal gating and can result in useful signal reception before the onset of the earliest room reflections. We will discuss this topic in a later section.

13.7 An Overview of Transform Measurement Methods

The modern personal computer (PC) is the heart of most of the transform testing methods used today. As opposed to the analog techniques we have discussed in earlier sections, the PC stores the results of the measurements, allowing them to be processed later and graphically reformatted as desired. By comparison, the output of the earlier analog testing methods was a piece of paper—a graph. If other formatting was desired, the measurement had to be rerun.

The Fourier transform is the basis of these new methods, and it provides a mathematical connection between time and frequency domains. Facilitated by the Fast Fourier Transform (FFT) algorithm, the modern PC can process the data very nearly on a real-time basis.

Fundamentally, transform systems use a test signal that is specific in both amplitude and phase over the relevant bandwidth. For example, the unit impulse function at some reference $t = 0$ s is equivalent to an unbounded number of individual cosine wave signals whose values are all maximum at $t = 0$, so both amplitude and time (phase) relationships are known for every frequency component.

The unit impulse function sounds like a very sharp "snap." Such a signal can be used to drive a system, and the measurement of that system directly compared with the driving signal. From this data, the transfer characteristics of the device under test can be calculated and displayed in a variety of ways. The major drawback with the impulse test signal is that it has a very high peak to rms ratio (crest factor) and may cause system overload if any attempt is made to secure a good signal-to-noise ratio.

Some systems use a linear frequency sweep as the test signal. The relation between the linear sweep and the impulse function is a fundamental one; if the impulse function is passed through an all-pass network that linearly shifts phase from HF to LF, the resulting output will in fact be a downward-gliding linear frequency sweep. The swept frequency has an advantage in that it can be fed to the system under test at a fairly high level, with little tendency for overload, thus ensuring a very high signal-to-noise ratio. Obviously, a certain amount of time is required to make the frequency sweep.

Maximum length sequence (MLS) systems use a test signal that is a pseudorandom sequence of on-off pulses that repeat at fixed intervals, with a sequence length defined by the relationship $2^N - 1$, where N is any integer. The autocorrelation of the signal will be unity at the sequence length, but when compared with the return signal via the measurement microphone, the crosscorrelation, as a complex

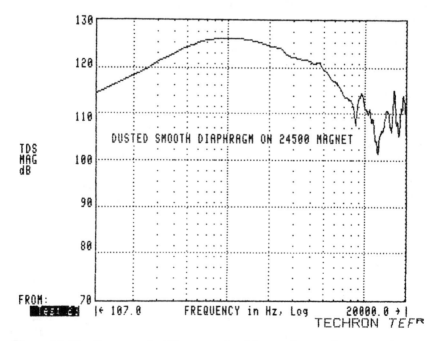

Figure I3-I3. Response of a HF compression driver made on a plane wave tube using the TEF measuring method. (Data courtesy JBL, Inc., and Fancher Murray.)

value, will contain data that describes the transfer function of the system under test. Without going into the further nature of the signal, we will say only that it provides an excellent signal-to-noise ratio and sounds very much to the ear like a continuous white noise (equal energy per cycle) signal.

All of the transform systems require high signal-to-noise ratios and low nonlinear distortion in the device under test if the resulting frequency response data is to be useful.

13.7.1 Examples of Transform Measurements

Traditional frequency response measurements can be easily duplicated by transform methods. Figure 13-13 shows a measurement of a HF compression driver made on a PWT using the Techron TEF (Time-Energy-Frequency) system. When all parameters are matched, the curve virtually overlays similar measurements made on analog chart recorders.

The curves showing horn-driver complex impedance data, both modulus and phase angle, shown in Figure 13-14, were made using the MLSSA maximum-

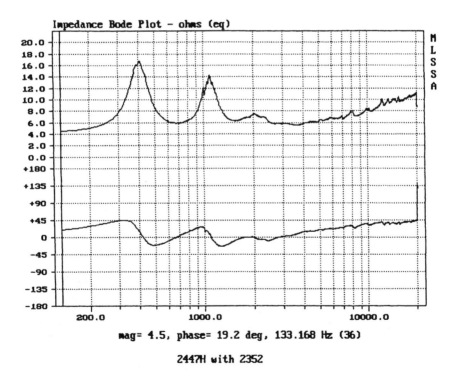

Figure 13-14. Loudspeaker modulus and phase of impedance made in free space using the MLSSA method. (Data courtesy JBL and Allan Devantier.)

length sequence system, with driver voltage and current values being fed to the analyzer.

13.7.2 Time and Frequency Trade-offs

Transform methods come into their own when used to generate a family of time-frequency curves on a single graph. Hidden lines in the display make it possible to present realistic and intuitively obvious three-dimensional displays.

For example, the data shown in Figure 13-15 shows successive level versus frequency plots over discrete time intervals. In this set of graphs, level is measured along the vertical scale, frequency along the left-to-right scale, and time along the back-to-front scale. This family of curves is normally known as a time-energy-frequency (TEF) or "waterfall" display. This display shows details of stored energy and transient overhang in acoustical systems, and as such the displays are valuable in both loudspeaker and room analysis.

There is, however, a pitfall associated with these measurements which has to do with joint period-frequency resolution. The relation between period and frequency is given by:

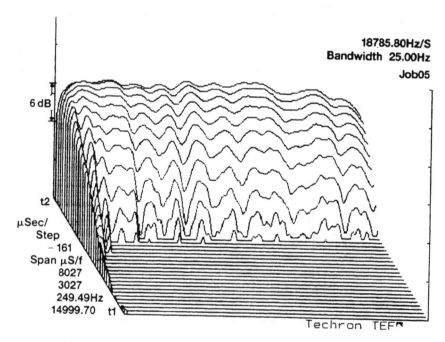

Figure 13-15. A "waterfall" display showing successive relationships among time, frequency, and level in the decay pattern of a loudspeaker, using the TEF measuring method.

$$\text{Period} = 1/\text{frequency} \qquad (13.1)$$

which implies that the product of period and frequency is always unity.

Examples of this tradeoff can be seen in Figure 13-16. In the family of successive time curves shown in Figure 13-16*a*, there is considerable LF resolution, but time data shows relatively little change between successive "slices". In order to see more accurate time data we have to give up LF resolution, as shown in Figure 13-16*b*. Both presentations give reliable information, but the engineer

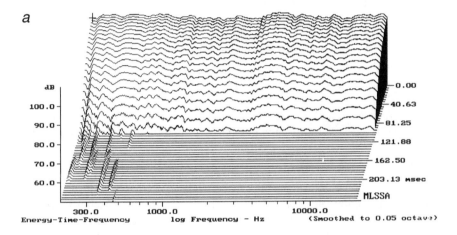

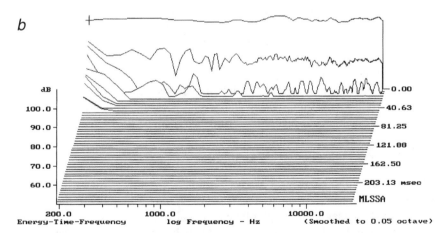

Figure 13-16. Illustration of the time-frequency trade-off. Time intervals spaced so that considerable LF detail can be seen (*a*); increasing time resolution diminishes the LF detail but presents more time information (*b*). (Data courtesy JBL and Allan Devantier.)

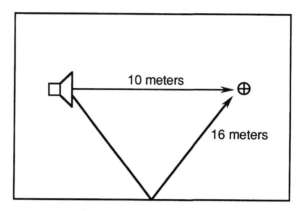

Figure 13-17. Frequency response made in a room using gating techniques. Return signal gated off at 6/344 s, corresponding to one period for 58 Hz.

must know what to look for. A great deal of excess and confusing data has been published in recent years through ignorance of these fundamental relationships.

13.7.3 Gating Techniques

Frequency response measurements can be made in normal acoustical spaces if the signal reception can be gated off before the first room reflections arrive at the measurement microphones. It is of great benefit to be able to make in situ frequency response measurements, but LF resolution is dependent on the time interval between reception of direct sound and that of the first reflection. Considering the details of Figure 13-17, the distance between the loudspeaker and microphone is 10 m, and the total distance traveled by the first reflection is 16 m. The difference here is 6 m, representing a delay of 6/344 s. This is the period corresponding of a frequency of 344/6, or 58 Hz, which would certainly be adequate for many large-room acoustical measurements.

Typical home environments are much smaller, and the delay times much shorter, corresponding to higher useful frequency limits for LF assessment.

13.8 Optical Measurement Techniques

The transducer design engineer always has a need for direct visual examination of cone and diaphragm motions. The simplest method here is to use the stroboscope. The best analogy to suggest here is watching a western movie and seeing the stagecoach wheels come to a halt, and slowly start to spin backward, even as the stagecoach actually increases in speed.

The rotation of the wheel spokes is cyclic and is sampled by the motion picture camera at a rate of 24 frames/s. When the successive "spoke rate" corresponds to the frame rate, the spokes appear stationary.

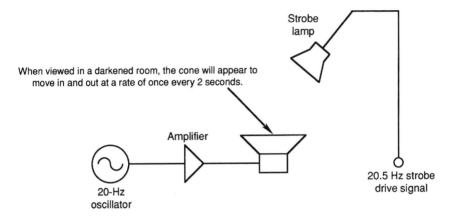

Figure 13-18. Principle of stroboscope testing of large cone motions of the driver.

The same principle can be applied to a loudspeaker as it is observed under a strobe light, as shown in Figure 13-18. The strobe light is a gas discharge device and as such can generate a very short pulse of light at rapid rates. In the example shown here, the driver is operated at 20 Hz, and the strobe light is is flashed at a rate of 20.5 Hz. When examined in a darkened room, the cone will appear to move in and out at a frequency of 0.5 s, or once every 2 s. Any gross departures from normal linearity will be visible to the naked eye.

Various kinds of laser measurements can be applied to loudspeaker analysis, when the cone is moving at microscopic displacements. Here, a laser beam is scanned over the cone or diaphragm and the reflections measured, stored, and displayed in a variety of ways. Wavelength interferometry can be used, or Doppler shifts in the laser wavelength can be measured. As with the stroboscopic method, time-varying behavior can be displayed with considerable amplification. A snapshot example of this technique is shown in Figure 13-19 for a 100-mm compression driver dome radiator operating at several frequencies.

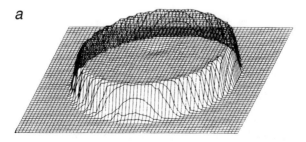

Figure 13-19. Laser techniques showing diaphragm motions at frequencies of 1 kHz (*a*); 4 kHz (*b*); and 8 kHz (*c*). (Data courtesy JBL, Inc. and Fancher Murray.)

b

c

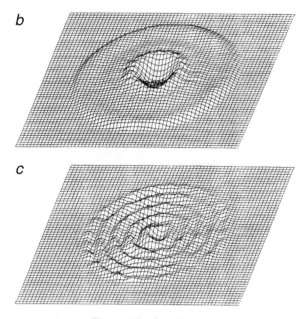

Figure 13-19. *Continued*

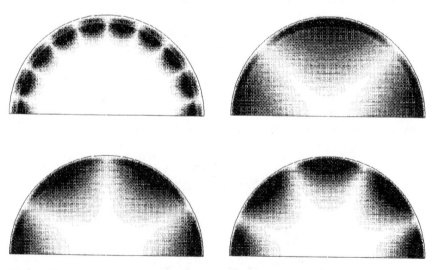

Figure 13-20. Finite element analysis. Modeling of the performance of a 100-mm diaphragm at several high frequencies. (Data courtesy JBL, Inc.)

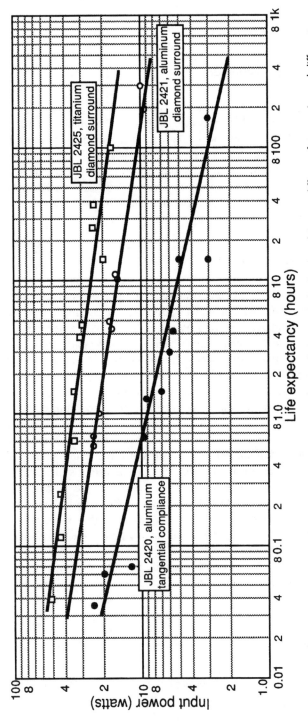

Figure 13-21. Failure modes. Cumulative failure data for a sample of aluminum diaphragms driven at different frequencies and different powers. (Data courtesy JBL, Inc. and Fancher Murray.)

13.9 Modeling Techniques

If sufficient performance data on a physical system can be measured and analyzed, it is possible to model the physical process for estimating actual performance. As we saw earlier in this book, the nonlinearities of magnetic circuits can be modeled by finite element analysis (FEA). These techniques can be extended to cones and domes. An example is shown in Figure 13-20 which shows a 100-mm diaphragm used in a compression driver. The various radial and circumferential modes are clearly seen and are of much help to the transducer design engineer in further stages of product refinement.

13.10 Destructive Testing

While we normally think of destructive testing in terms of automobiles, housing units, and the like, having to do with structures concerning human safety, destructive testing techniques are very useful in loudspeaker assessment. Loudspeaker drivers are routinely tested to the breaking point, primarily as a means of quality assurance and of assessing manufacturing control of all processes. This concern also extends to the design stages of HF compression drivers where the moving systems are normally made of materials that exhibit stress related failure modes.

Under typical stress-strain (force-displacement) conditions, aluminum, for example, will exhibit a unique failure history over long operating periods. For small diaphragm excursions, many flexures may be endured before weakness, and eventual failure, sets in. For larger excursions, the failure point will be reached with fewer flexures, or at an earlier time in the driver's history.

Many plastics do not exhibit such a "bend and break" behavior. Metals do exhibit this tendency, but to differing degrees. Aluminum exhibits relatively high fatigue buildup, while titanium and beryllium exhibit respectively less. Any of the metal materials may shatter or split at any time if specific displacement limits are exceeded.

The failure rate of any material can only be inferred from statistical observation of many samples, and an example of the behavior of aluminum is shown in Figure 13-21.

Bibliography

Collums, M., *High Performance Loudspeakers*, Wiley, New York (1991).

Davis, D., and Davis, C., *Sound System Engineering*, Sams, Indianapolis, IN (1987).

Heyser, R. C., "Acoustical Measurements by Time Delay Spectrometry," *J. Audio Engineering Society*, Vol. 15, No. 10 (1967).

Murray, F., "MTF (Modulation Transfer Function) as a Tool in Transducer Selection," in *Proceedings of the Audio Engineering Society 6th International Conference*, May 1988.

Rife, D., and Vanderkooy, J., "Transfer-Function Measurements with Maximum-Length-Sequences," *J. Audio Engineering Society*, Vol. 37, No. 6 (1989).

Vanderkooy, J., "Aspects of MLS Measuring Systems," *J. Audio Engineering Society*, Vol. 42, No. 4 (1994).

CHAPTER

14

Loudspeaker Specifications for Professional Applications

14.1 Introduction

In this chapter we will discuss the principal specifications of loudspeaker components used in professional sound design. These include:

1. On-axis frequency response
2. Impedance
3. Sensitivity
4. Electrical power input ratings (thermal and displacement)
5. Power compression
6. Distortion (discrete harmonics)
7. Directional characteristics
8. Thiele-Small parameters

The topics covered here are central to specification writing and proper system layout, and it is essential that they be clearly understood.

14.2 On-Axis Frequency Response

The familiar on-axis frequency response curve of a transducer or system is normally referred to a distance of 1 m with a stated input power or voltage. Most commonly used is an input of 1 W or an applied voltage of 2.83 V rms. The curve is often run at a distance greater than 1 m, and the results are adjusted for an equivalent 1 m distance.

In the listing of specifications for a device or a system, tolerances on the response may be stated. For example, the usable frequency range of a device may be given as the nominal the range over which the response is no lower than

−10 dB, relative to the midband rated sensitivity, but engineers are probably more interested in the range over which the response is ± 3 dB. The primary data is shown in graphical form as a function of level versus frequency. For systems, the range of any of the system's level controls is normally indicated on the on-axis response graph. An example of such a graph is shown in Figure 10-6c.

14.3 Impedance

A plot of the impedance modulus may be presented on the same graph as the on-axis frequency response. The modulus varies with frequency, and it is of special importance to the system designer to know the lowest value of impedance that a device will present to an amplifier. Care should be taken by the specifying engineer to consider the effects of paralleling of loads. As a rule, a plot of the impedance phase response is not given; rather it is useful to state at what frequencies the phase angle may be in excess of ± 60°.

14.4 Reference Sensitivity Ratings

A loudspeaker system or driver is normally given a sensitivity rating based on a band-limited pink noise power input of 1 W measuring the L_p at some distance along the principal axis of the device. The reference distance for the measurement is always 1 m. The voltage, E, required to produce 1 W input is derived from the nominal impedance, Z, of the device by the following equation:

$$E = \sqrt{Z}$$

Thus:

Nominal impedance (Ω)	Applied voltage (V rms)
16 Ω	4
8 Ω	2.83
4 Ω	2 Vrms

As we have noted, the impedance of a system or a transducer is not constant across the frequency band, and the term "nominal impedance" describes an average value that is approximately 15% greater than the minimum value of impedance the device exhibits. Most drivers and systems are conveniently designed around nominal impedances of 4, 8, and 16 Ω.

The signal bandwidth used for sensitivity measurements on systems is normally 500 Hz–2.5 kHz. For LF drivers and systems, the noise bandwidth is usually 200–500 Hz, and for HF systems it is usually 1–4 kHz.

In older specification data we may find references to 1 W sensitivity measured at a distance of 4 ft. This measurement will be 1.7 dB lower than the corresponding 1 m measurement.

Another rarely encountered standard is the old Electronic Industries Association (EIA) sensitivity rating of 1 mW sensitivity at a reference distance of 30 ft. This value will be 49.2 dB lower than the 1 m, 1-W rating.

For professional system design it is vital that the sensitivity data be accurate, since a small error, in a large sound system installation, can add up to many dollars spent unnecessarily on amplifiers—or worse, an inadequate specification of required power. Another potential problem is the increasing tendency of some manufacturers to use voltage sensitivity measurements. Here, a 2.83 V rms signal is used to make the sensitivity measurement. If the device under test has a nominal impedance of 8 Ω, the measurement is equivalent to an input of 1 W. However, if the device has an impedance of 4 Ω, it will receive 2 W from the source and generate an output 3 dB greater than for 1-W input. The picture can get very confusing, and users are urged to make careful calculations of their own. If the impedance of the device is not stated, and a voltage sensitivity is given, the data is virtually useless.

The boundary conditions adjacent to transducers or systems play an important role in their LF measured frequency response and directivity properties. For example, most HF devices are measured in a free field; that is, the horn is freestanding in space, and there are no nearby surfaces that could influence sound propagation from the mouth of the horn. By comparison, LF systems are normally measured mounted in a large, solid boundary out-of-doors. We refer to the former as full-space mounting, or 4π; the latter is called half-space, or 2π mounting. In 2π mounting, the adjacent surface provides a mirror image of the transducer that will reinforce its LF output by 3 dB, while providing a minimum DI of 3 dB. This results in a net increase of 6 dB at low frequencies as compared to free-field mounting of LF systems.

14.4.1 Sensitivity Data for Compression Drivers

There are no rigorous standards for compression driver sensitivity measurements, and the reader of specifications will have to look closely if driver comparisons between manufacturers are to be made. However, for drivers mounted on horns, the same 1 W, 1 m method we have discussed will apply.

Driver response is customarily measured on a PWT with a modest power input of 1 mW. The cross-sectional area of the tube will then be a factor in the measured pressure level in the tube. For many years it was more or less standard practice to adjust the pressure level reading to match that of a 25-mm (1 in.)-diameter PWT, whether or not the driver had a 25-mm exit diameter. As driver manufacturers proliferated, and models of different exit diameters were introduced, a certain amount of confusion was bound to set in.

The data shown in Figure 14-1*a* will enable the reader to determine the efficiency of a given driver, based on the diameter of the PWT and the resulting pressure level for an input of 1 mW. Of course, it is their relative efficiency that will facilitate the driver-to-driver comparisons.

To complicate matters even more, there is a movement in some quarters to establish a reference intensity in PWT measurements of 1 mW/cm^2. Comparison data for this approach is shown in Figure 14-1*b*.

14.5. Power Ratings of Drivers and Systems

Assigning power ratings to transducers and systems is a complicated matter. What we are attempting to do is give the user a proper guideline for choosing a power amplifier that is large enough so that its output signal will not clip during normal use—but not so large that we run the risk of burning out the loudspeaker or mechanically damaging it.

These two requirements are not always easy to sort out. For example, a typical three-way cone-dome monitor loudspeaker may carry a nominal input power rating of 150 W, using a pink noise signal. Examining this specification closely, we note that the noise signal generally has a specified crest factor of 6 dB. This means that the average power delivered to the loudspeaker will be 150 W, but instantaneous peak power values of four times (+ 6 dB) that amount, or 600 W, will be delivered to the system.

It has been noted that single-frequency signals in the 300–800 Hz range of 1200 W can be safely delivered to the system—if such signals are short enough in duration and there is a duty cycle (power on/power off) that allows the voice coil to cool down before the next burst of input power. Also, it has been determined that sustained operation at 10 Hz at a power input of only 75 W can mechanically harm the LF driver due to excessive displacement of the cone.

Obviously, a manufacturer cannot consider all of these extreme contingencies when setting power ratings, so practical ratings must be based on normal operation of systems according to good engineering practice. It is further assumed that the user will not intentionally abuse the system.

For full-range loudspeaker systems, most professional manufacturers use a shaped pink noise (6 dB crest factor) spectrum established by the IEC. This spectrum approximates modern music and speech spectra fairly well, and the 6 dB crest factor implies that signal peaks 6 dB higher can be safely handled if they are within the operating bandwidth of the system. The stated rating is the highest average power input that the system can sustain for a long period of time over the specified bandwidth without any sign of permanent damage. This is essentially a thermal rating; that is, if the loudspeaker system fails in this test, it is normally as a result of overheating, with consequent burnout of the voice coil.

Figure 14-2 shows weighting curves that are used in loudspeaker system evaluation. Curve 1 in Figure 14-2*a* has been suggested by the EIA for loudspeaker

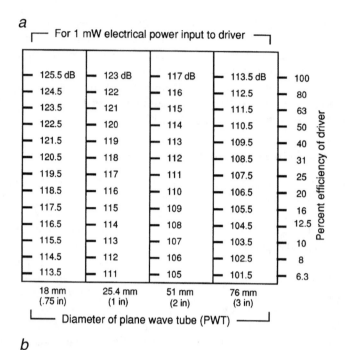

Figure 14-1. Compression driver sensitivity. Constant 1 mW input data (*a*); constant intensity in the PWT (*b*).

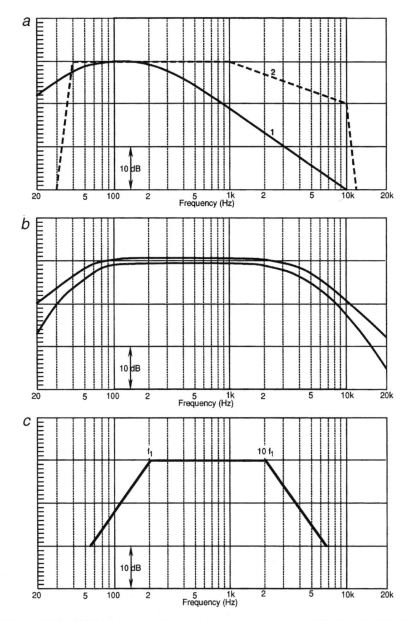

Figure 14-2. Weighting curves for loudspeaker power testing. EIA data (*a*); IEC data (*b*); AES data (*c*).

system measurement, with white noise as the input. Curve 2 in Figure 14-2*a* is an updated curve as suggested by the EIA for modern program material.

The curve shown in Figure 14-2*b* is suggested by the IEC for loudspeaker system measurements, with a signal input of pink noise with a 6 dB crest factor.

The curve shown in Figure 14-2*c* is recommended by the AES for rating individual LF components. Low-frequency drivers are normally power rated according to the AES method. In this measurement, a one-decade (10-to-1) frequency range of band-limited pink noise (6 dB crest factor) is applied to the device under test. The manufacturer states the particular frequency decade over which the tests have been run, and the test is run with the transducer placed in free space (not in an enclosure). This procedure is an attempt to qualify a given driver by itself, with no necessary relationship to the enclosure it will be mounted in.

If the frequency decade is chosen at higher frequencies, the rating will primarily reflect the thermal limit of the device. If the decade extends low enough in frequency, it will reflect, to a greater or lesser extent, the displacement limits of the device.

High-frequency drivers are similarly rated by choosing a particular frequency decade with a given horn load on the driver.

It is up to each manufacturer to select and state the appropriate frequency decade for a given transducer, and it is not uncommon for a transducer to carry more than one AES rating, depending on the frequency decades that have been chosen.

14.6 Power Compression

When high power is delivered to a transducer over relatively long periods of time, the voice coil will heat up and its resistance will rise. As a result, the loudspeaker will draw less power from the amplifier and its sensitivity will decrease. A typical listing is shown here for a professional 380-mm-diameter driver with a 100-mm voice coil:

At −10 dB power (60 W): 0.7 dB
At −3 dB power (300 W): 2.5 dB
At rated power (600 W):   4.0 dB

These values assume that the device under test has reached thermal equilibrium at each of the power values listed. Additional data on power compression is presented in Chapter 9.

Not all manufacturers routinely present compression data in their specification sheets, and there are no generally agreed upon value limits.

14.7 Distortion

The standard method for showing distortion is to plot second- and third-harmonic components on the same graph with the fundamental. The distortion curves may be raised 20 dB for ease in reading, and the normal power input for the measurements is usually one-tenth (−10 dB) rated power. An example of this method is shown in Figure 2-25.

At any point on the graph where the fundamental and distortion curves intersect, the value of distortion is 10%. Because of the 20 dB offset of the distortion curves, the actual value of distortion is 20 dB lower than the fundamental, corresponding to 10%. In this figure, the value of third-harmonic distortion at 40 Hz is 10%.

14.8 Directivity Performance of Systems and Components

As discussed in several preceding chapters, directional data on loudspeakers covers much ground, and Figure 13-9 summarizes the chief graphical methods of display. It is appropriate in this chapter to suggest data presentation methods and performance guidelines for various areas of professional sound work.

14.8.1 Speech Reinforcement

The major problem in selecting components for speech reinforcement in reverberant spaces is that of ensuring proper directivity control throughout the speech range. Traditionally this range has been assumed to be from 500 Hz to about 4 kHz, but many in the field have always felt that systems sounded better when good pattern control was maintained down to 200 Hz and up to 8 kHz.

Most HF horns, large or small, will provide good loading at least an octave below the point at which pattern control in the narrow plane starts to widen. Assuming that the HF driver can handle the anticipated load, many designers may opt for a lower crossover frequency in an effort not to carry the LF portion of the system too far beyond its range of flat power output capability. Good engineering calls for an innovative approach in the upper bass region; for example, the use of a vertical array of relatively small LF drivers to fill in that portion of the spectrum with good forward directivity.

While this is not actually a driver design problem, the vast majority of ceiling distributed arrays in public places suffer from insufficient coverage at frequencies up to 2 kHz. The present state of the art in small ceiling drivers that are economical to build permits a nominal radiation cone of about 90° (−6 dB) at 2 kHz. The data presented in Figures 11-6 and 11-7 demonstrates the need for a fairly dense array. If this is not possible for whatever reason, the system will probably be better off with a small number of wide-angle radiating devices, such as the Omnisphere, hanging from the ceiling at calculated locations.

14.8.2 Music Applications

In the recording studio it is essential to achieve good horizontal coverage for both seated and standing personnel. The vertical coverage at high frequencies may be allowed to vary from relatively wide in the mid-range to about 45° at the highest frequencies. The relatively dry studio acoustic environment will minimize the broad shift in actual power response of the system. High-frequency elements exhibiting wide variations in directional response should be avoided.

14.9 Thiele-Small Parameters

The T-S parameters have become an important specification for any professional LF driver. They should be accurately determined from an adequate sampling of actual production units. Particular attention should be paid to X_{MAX} and P_E, since these parameters determine the large-signal performance of the driver. Normal production variations in cone mass should be negligible, and incoming inspection procedures should pick up any shift here.

The stiffness of surrounds and spiders may tend to vary to some degree, causing the V_{AS} and Q_{TS} parameters to shift. The main question here may not be the variation itself, but the effect that variation may have on system performance. In most ported systems, the LF alignment is dominated by the enclosure parameters, and slight shifts in these parameters may have little practical effect. Nevertheless, quality assurance procedures in the manufacturing plant should take account of possible variation in these parameters.

Bibliography

"Recommended Practices for Specifications of Loudspeaker Components Used in Professional Audio and Sound Reinforcement (AES 2-1984; ANSI S4.26-1984)," *J. Audio Engineering Society*, Vol. 32, No. 10 (October 1984).

Aspects of the Home Listening Environment

15.1 Introduction

While the professional sound world relies heavily on loudspeaker specifications in systems design and layout, the choice of loudspeakers for home listening is likely to be the result of a far more personal set of experiences and associations. This can be inferred from a casual reading of high-fidelity system manufacturers' literature and advertisements.

But there is nothing casual about a truly first-rate home audio system. When care has been given to the choice of loudspeakers, room treatment, and electronic componentry, a consumer playback system can outperform a more perfunctorily designed professional system. And often for a lesser price.

A big obstacle to putting together a truly fine home system is the dedication of space required for it. In a family setting this may be difficult to justify, and perhaps as a result, many high-fidelity loudspeaker companies have put significant efforts into the conceptual design and engineering of smaller systems capable of excellent performance in the modern home environment.

In this chapter we will examine many of these concerns, stressing the nature of the listening room as a very complex transmission path between loudspeaker and listener.

15.2 Listening Room Boundary Conditions: the Laboratory Meets the Real World

Virtually all loudspeaker design is carried out in simulated open space (4π) or ground plane half-space (2π) environments. The professional designer can infer much from this data, but the consumer, and the consumer dealer, are often left in the dark. The actual home loudspeaker environment is apt to be somewhere between a 4π and 2π condition, and some loudspeakers, notably corner horns, are intended to be operated virtually in a $\pi/2$ (one-eighth space) environment.

Roy Allison (1974) is one of only a handful of loudspeaker designer-manufactur-
ers who have tackled this problem headlong in making specific loudspeaker
placement recommendations for the consumer.

As shown in earlier chapters, halving the solid angle into which an omnidirectional device radiates will double its efficiency. Additional power doublings take
place at adjacent pairs of surfaces (dihedral corners) and at adjacent trios of
surfaces (trihedral corners). The precise nature of this effect is wavelength dependent, as Allison shows in Figure 15-1. The curves show the normalized power
output of an omnidirectional source as a function of distance (x, y, or z) and
signal wavelength (λ), a wall-floor boundary (B), and a corner-floor boundary (C).

Considering only curve A of Figure 15-1, for a loudspeaker in the far field
(beyond $x/\lambda = 2$), the output has been assigned an arbitrary reference level of 0
dB. As the loudspeaker is moved closer to the boundary, there will be a slight
ripple in the response due to reflections from the boundary. This ripple is wavelength dependent, and at very long wavelengths ($x/\lambda = 0.02$) the reflection will
be virtually in-phase with the primary source.

Curves B and C of Figure 15-1 show similar data for sources of a given

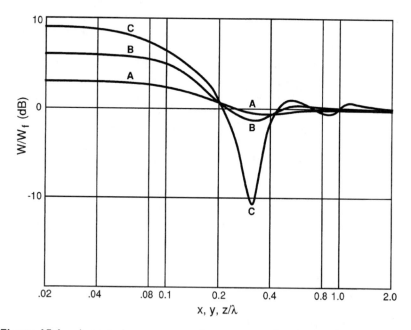

Figure 15-1. Acoustical power output for a source relative to its free-field output
when located next to a single wall (curve A), two right-angle walls (curve B), and two
right-angle walls and a floor (curve C). The horizontal scale shows the source location
in terms of wavelength (x/λ, y/λ, and z/λ). For two- and three-boundary cases, the curves
apply only on lines of symmetry. (Data after Allison, 1974.)

distance and wavelength from a pair of perpendicular surfaces and a trio of perpendicular surfaces. In these cases the response irregularities are more pronounced, due to the multiple out-of-phase reflected images.

The picture is a complex one, and Allison has worked out several examples, as shown in Figure 15-2. In general, the closer a LF driver can be brought to a boundary, or set of boundaries, the smoother the response will be. If the driver cannot be operated in this position, it may be better overall if it can be placed well out into the listening space on a stand. In this case the reflected images will be farther removed from the listener and the response thus made smoother.

15.3 Room Modes

The foregoing analysis assumed that the boundaries adjacent to the loudspeaker were quite large. In typical rooms, the conditions discussed by Allison will be modified by discrete frequencies, or room modes, at which the room will become very responsive. Extending the analysis of boundary conditions to an arbitrary space is complicated, but if we can limit the discussion to a rectangular space the analysis will be much simpler.

The normal modes of a rectangular room are given by the following equation:

$$f = \left(\frac{c}{2}\right) \sqrt{(n_l/l)^2 + (n_w/w)^2 + (n_h/h)^2} \qquad (15.1)$$

where c is the speed of sound, l, w, and h the length, width, and height of the room, respectively, and the three values of n taken separately as integer values.

At middle and high frequencies in all but the smallest spaces, room modes will overlap, and the room will respond to a loudspeaker more less uniformly with a pattern of discrete reflections that are normally described statistically, leading to a definition of reverberation, or "ring-out," in the space.

At lower frequencies the modes are not so closely spaced and the acoustical description of the room changes significantly; the natural modes of the room set up a variation in level over space that profoundly affects the transmission of sound from loudspeaker to listener. While there is certainly sound propagation directly from source to listener at frequencies other than the natural modes, such propagation will be relatively low in level.

In many underdamped rooms, certain modes may be prominent enough in the 100–250 Hz range to cause audible coloration of normal speech sounds produced in the space. If this situation exists, it is a clear indication that sound reproduction in the space will suffer, and every effort should be made to correct it before attempting to install a stereophonic listening system. This problem is most often found in smallish rooms with hard structural surfaces, such as concrete floor slabs, cinder block walls, and the like. Flexible surfaces, such as dry-wall construc-

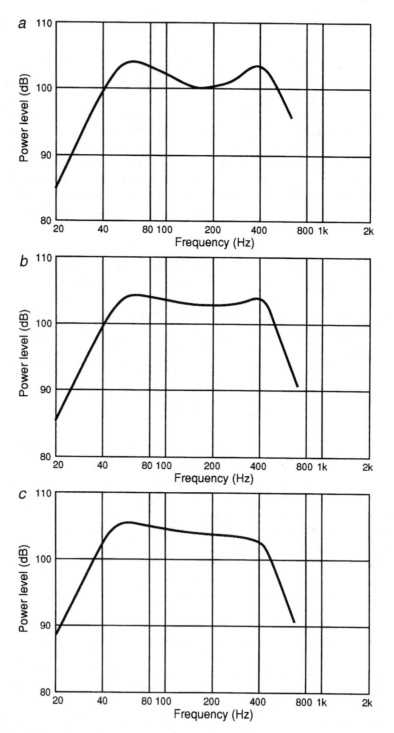

Figure 15-2. Location for a single LF radiator. Back of enclosure located 1 in. from wall (a); LF driver location 0.5 in. from wall (b); LF driver location 0.5 in. from wall and floor (c). Data after Allison, 1974.)

tion and large expanses of glass will alleviate this specific problem—but will usually require drapery or other wall treatment to minimize MF and HF reflections.

In most rooms, the LF modes along the listening axis will predominate, and their pressure distribution will be as shown in Figure 15-3a–d. Only the first four or so modes may be significant in determining the relative levels at the listener. Where both source and listener are high on a given modal response curve, the coupling between the two will be strong for that mode. When a loudspeaker is placed at a null point for a particular mode, the coupling will be weak. However, if a dipole loudspeaker is placed at a null point along the curve and oriented along the direction of the mode, the coupling will again be strong.

As an example, consider a listening room that has floor dimensions of 4 and 6 m. Let the listening axis be along the 6 m dimension of the room. Accordingly, we will use n_l values of 1, 2, 3, and 4, setting all n_w and n_h values to zero. These values are individually entered into Equation (15.1), giving frequency values of: 28.7, 57.3, 86, and 115 Hz, respectively.

The nature of room modes is that they are the preferred modes of the room itself, and the response of a loudspeaker, if appropriately positioned, will excite room resonances at those frequencies. Those frequencies will generally be louder to the listener than nonmodal frequencies, but the distribution will vary considerably throughout the room.

As a practical matter, it would be desirable to excite the room at the lowest room mode (28.7 Hz) in order achieve added room support at that low frequency. Examining Figure 15-3a and looking at the pressure distribution for $n = 1$, we see that both listener and loudspeaker should be placed so that each is fairly close to the ends of the room. Moving the listening and loudspeaker positions

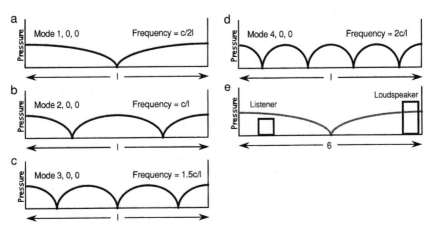

Figure 15-3. Pressure distribution for first four axial modes in a rectangular space along largest dimension (*a–d*). Adjusting loudspeaker and listening positions for spectral balance of the lowest axial mode at 28.7 Hz (e).

farther from or closer to the walls will determine the spectral balance balance precisely, as shown in Figure 15-3*e*. As a practical matter, it would be best to move both loudspeaker and listening position slightly into the room in order to reduce the coupling between the two in order to achieve the best overall acoustical balance. However, we should not forget Allison's recommendations regarding the spacing of the LF driver and its closest room boundaries.

15.4 Room Treatments at Mid- and High Frequencies

In general, a home listening space is governed by boundary reflections and mode structure below about 250 Hz. Above that frequency the mode structure gradually becomes fairly dense, and the response at the listener for frequencies higher than about 500 Hz is dominated by the specific directional properties of the loudspeaker and the MF and HF reverberant nature of the room. The range between these two frequencies is a transition zone between the two modes of operation.

Stated somewhat differently, at low frequencies the response is room dominated, while at high frequencies it is loudspeaker dominated. For a given direct field level, a more highly directional loudspeaker will produce less reflected acoustical power in the listening space; however, care should be taken that the room is not so heavily damped that its acoustic character seems oppressive to the listener. Discrete reflections, especially from the sides, should be minimized, and wall surfaces that provide diffusion may be better than those that absorb sound power. Irregular and reticulated surfaces are excellent in this regard, and the surface details shown in Figure 15-4 are useful in listening room treatment. The quadratic residue diffuser in particular affords uniform diffusion over a relatively wide frequency range (Schroeder, 1986).

Large exposed areas of floor, wall, and glass surfaces should be covered at least in part by carpet, adjustable drapery, and absorptive or diffusive wall treatments. The option of wall-to-wall carpeting should be considered with caution, inasmuch as it may not actually be necessary for the control of room reflections.

Examples of location and room treatment for two kinds of loudspeakers are shown in Figure 15-5. The system shown in Figure 15-5*a* is a traditional four-way cone-dome design with a characteristic DI plot that varies from 0 dB at low frequencies to about 14 dB at the highest frequencies.

The first consideration is to locate the loudspeakers so that LF response is smooth and free of peaks. If the room is well damped, with generous amounts of LF absorption, the job may be fairly easy. While Allison's recommendations are always appropriate, they are specific to loudspeaker systems whose LF drivers can be positioned as required without compromising midrange and HF projection into the room. In the room plan view shown at the left, dimensions A and B should be experimented with at some length. If both A and B are reduced to small values, then there will be a considerable LF rise, as is seen in Figure 15-

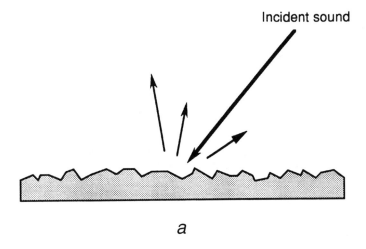

a

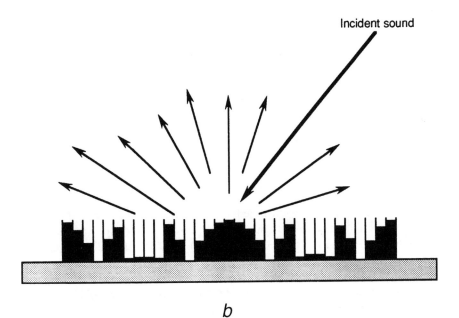

b

Figure 15-4. Sound diffusion from random reticulated surface (*a*) and from quadratic residue refracting surface (*b*).

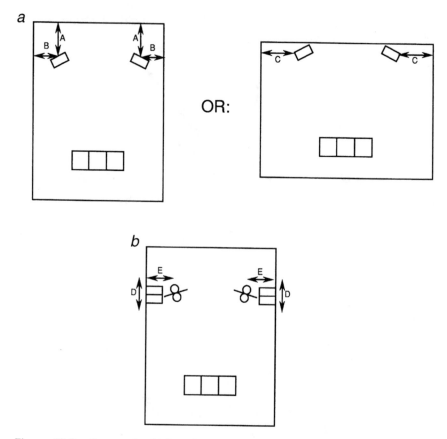

Figure 15-5. Suggested guidelines for room placement for a cone-dome system (*a*); for a dipole system (*b*).

1. If A and B are increased so that the loudspeakers are well out from the walls, there may be a smooth LF rolloff below, say, 125 Hz. A and B can be experimented with individually, and the ideal (or acceptable) balance determined by ear. A final adjustment of the seating position should then be made, primarily in order to adjust the stereo listening angle. Once this has been accomplished it is a relatively easy matter to finish the room in terms of midrange and HF absorption, largely to the taste of the user.

Some users may prefer the room plan view shown at the right. Here, the loudspeakers have been placed along the longer wall and as such will be well inboard of the adjacent corners. In this case, loudspeaker positioning fairly close to the wall may be best, with distances C adjusted for best listening angle.

The system shown in Figure 15-5*b* is of the dipole type. Many readers may assume that a dipole loudspeaker, because it radiates as much sound power from the back as it does from the front, unnecessarily contributes to the reverberant field in the listening room. This is not the case, inasmuch as the DI of the dipole never drops below a value of 4.8 dB. (Recall that the dipole has minimal radiation at off-axis angles of ±90°.)

The chief problem in interfacing the dipoles in a typical listening space is to ensure that there will be adequate LF response. It will be useful to identify those room modes of second and higher order along the listening axis of the room and experiment with placing the loudspeakers at nodal points, perhaps with the loudspeakers located fairly closely to the side walls to increase LF response. The difficulty here is that the room may be too wide for the normal stereo stage width of 45°–60°.

Some listeners prefer to augment dipole loudspeakers with sealed-cone LF systems—so-called monopole systems—in order to satisfy their bass requirements. Most devotees of dipoles, however, refuse to do this, preferring the natural bass rolloff as a simple price to pay for the special listening experience that these loudspeakers generally provide. Other designers have experimented with large area LF dipole arrays to augment the bass portion of the listening spectrum. If these devices are used, they should be placed against the long side walls of the room, with the full-range units inboard. Values of D and E may be experimented with for the best overall LF and MF balance, and the seating position adjusted for best listening angle.

In any event, once the LF coupling via room to listener has been satisfactorily made, the details of room treatment at mid- and high frequencies should not be materially different from the approach used with traditional cone-dome loudspeakers.

15.5 Optimizing Stereophonic Localization

In most listening environments there are a number of things the listener can do improve the quality of stereophonic imaging. If care has been taken in the basic setup, and if the room is acoustically balanced with regard to mid-LF and LF coloration, the imaging should be reasonably good to begin with. A good way to check for this is to set the system mode switch to mono, and then to play a pink noise signal over the system. There are a number of test discs for this purpose, and the proper signal to use here would be either a full range pink noise signal or one limited to the midrange portion of the spectrum. The listener should carefully assess the nature of the perceived signal while seated directly on the center line of the setup, equidistant from both loudspeakers.

The signal source should appear directly in front of the listener, located precisely on a line connecting the two loudspeakers. If all previous recommendations regarding electrical and acoustical balance have been made, then there should be no problems here. If the image appears to be vague or diffuse, the listener should recheck all matters of balance. If the problem persists it is likely a sign that the room is too live at mid- and low frequencies.

Considerable time should be spent in checking and refining image stability. If slight left-right nodding of the listener's head causes the image to wander widely from side to side, it is probably a sign that the loudspeakers are too widely placed. The normal stereo listening angle may range from 45 to 60°, depending on taste. When the subtended angle is in this range, normal side-to-side head movements should have little deleterious effect.

Increasing the seating range over which stereo imaging will be effective depends on the "cross firing" of the loudspeaker's primary axes. If it is desired to widen this area somewhat, the listener should experiment with toe-in of the loudspeakers so that their major axes cross slightly in front of the center listening position. It would be wise to position the loudspeakers rather precisely in doing

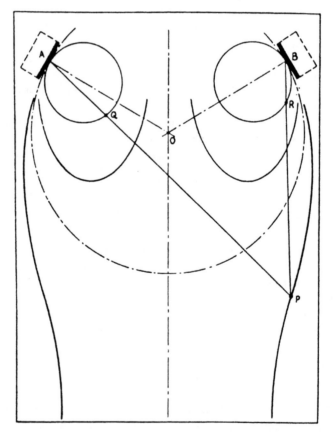

Figure 15-6. Bauer's dipole loudspeaker arrangement for extended stereo imaging. (Data courtesy J. Audio Engineering Society.)

this, since the trade-off between amplitude and time cues for off-axis listening is critical.

The best signal to use for making this assessment is the spoken voice, again with the system in monophonic mode. Finally, when the monophonic, or phantom center, image is well behaved, the listener can switch to the stereo mode and will probably be delighted with the results.

15.6 Loudspeaker Systems for Extended Imaging

Extending the notion of cross-firing stereo loudspeakers, several designers have come up with interesting systems concepts using asymmetrical radiation patterns or other patterns that exhibited a rapid falloff in response with respect to listening angle. Bauer (1960) examines the application of dipole loudspeakers cross-fired in front of the listening position. Since the dipole pattern is maintained well over the MF and LF ranges, a satisfactory trade-off between level and delay can be produced in the listening zone. Figure 15-6 shows Bauer's basic setup. The requirement of all these systems is to produce smooth off-axis response in addition to smooth on-axis response, something the dipole does naturally.

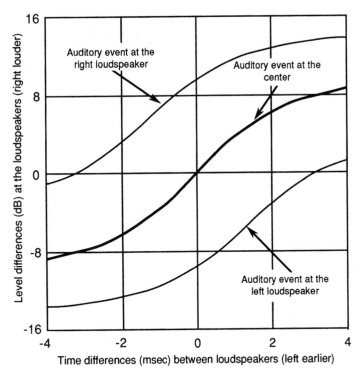

Figure 15-7. Franssen's data on center event localization.

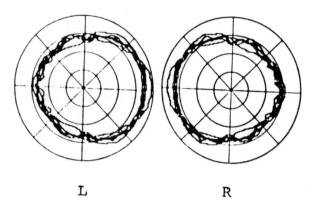

L   R

Figure 15-8. Details of Davis's extended imaging loudspeaker design. (Data after Davis, 1987.)

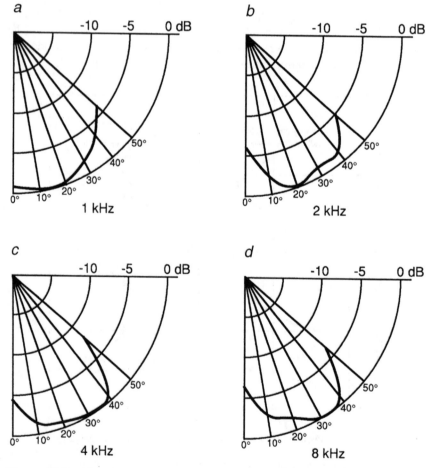

Figure 15-9. Use of asymmetrical horn patterns in extended stereo imaging. Polar graphs shown here are for the left channel; those for the right channel are mirror images. (Data courtesy JBL, Inc.)

Data regarding the quantitative aspects of time-amplitude trade-off is spotty, and the effect is very program dependent. As a starting point for discussions, we present Franssen's data in Figure 15-7. Although this data was generated for description of recording systems, it does enable the loudspeaker designer to target a range of operation for off-axis maintenance of a center image. The off-axis position is determined in milliseconds, and that value is entered into the graph. Then the required amplitude imbalance that will produce a center event is noted.

Davis (1987) described a rather complex cone-dome system that produced a consistent off-axis response that steered the center image appropriately. The principal lobes were aimed horizontally, as shown in Figure 15-8, and normal center listening took place along some off-axis angle. A listener to the right or left of that position moved into a range that was closer to the principal axis, and the increase in level provided a trade-off with the time cues.

Because asymmetrical horn radiation patterns are relatively easy to produce, they have also been used to produce systems for extended imaging. Eargle and Timbers (1986) describe a system using an asymmetrical HF horn to provide the necessary skewed response. Figure 15-9 shows the polar response of the system on several octave centers.

Bibliography

Allison, R., "The Influence of Room Boundaries on Loudspeaker Power Output," *J. Audio Engineering Society,* Vol. 22, No. 5 (June 1974).

Bauer, B., "Broadening the Area of Stereophonic Perception," *J. Audio Engineering Society,* Vol. 8, No. 2 (1960).

Collums, M., High Performance Loudspeakers, Wiley, New York (1991).

Davis, M., "Loudspeaker Systems with Optimized Wide Listening Area Imaging," *J. Audio Engineering Society*, Vol. 35, No. 11 (1987).

Eargle, J., and Timbers, G., "An Analysis of Some Off-Axis Stereo Localization Problems," Preprint No. 2390; presented at the 81st Audio Engineering Society Convention, Los Angeles (1986).

Schroeder, M., "Progress in Architectural Acoustics and Artificial Reverberation: Concert Hall Acoustics and Number Theory," *J. Audio Engineering Society*, Vol. 32, No. 4 (1984).

Toole, F., "Loudspeaker Measurements and Their Relationship to Listener Preferences: Part 2," *J. Audio Engineering Society*, Vol. 34, No. 5 (1986).

CHAPTER 6

A Survey of Exotic Transducers

16.1 Introduction

Standard dynamic and electrostatic transducers form the basis of mainstream loudspeaker system design. Indeed, there are few other principles of transduction that exhibit either the linearity or the acoustical output capability to form a basis for loudspeaker design. Nonetheless, there is always interest in transducers based on other physical principles or unusual variations of conventional technology.

There are a number of reasons why a manufacturer might want to develop a new transducer, among them:

1. An exotic driver, purely for its own sake, may have possible marketplace advantages.
2. Many exotic drivers are designed for 360° radiation in the horizontal plane, and this is considered an advantage by many systems designers.
3. In a few cases, an exotic driver has provided the basis for an entirely new wide-band systems approach, thus enabling the manufacturer to make a major design statement.

Some of the attendant problems are:

1. Difficulties in construction, which can lead to high manufacturing costs and production variations.
2. Some designs wear out in normal use and require periodic replacement of critical parts.
3. Many designs are inherently low in sensitivity and output capability, limiting the appeal of the products to a dedicated but small clientele.

In many cases, the technical and patent history of these devices is not clear, and detailed descriptions may not be easy to come by. Accordingly our coverage here will be broad and general.

16.2 Variations on a Magnetic Theme

All dynamic devices work on the principle of current flowing through an electrical conductor that is positioned perpendicular to a magnetic flux field. The resultant force produced on the conductor is mutually perpendicular to both the current flow and the magnetic flux. This relationship is fundamental to all dynamic transducers, and the useful acoustical output radiation from such device is normally along the same axis as the force that is produced by the driving mechanism.

16.2.1 The Heil Air Motion Transformer (AMT)

The air motion transformer (AMT), developed by Oskar Heil, represents a significant departure from the normal geometry of the dynamic driver. A perspective view of the AMT is shown in Figure 16-1a, and a horizontal section view of the diaphragm assembly is shown in Figure 16-1b.

The magnetic flux path is from left to right, and current flows through a flat conductor bonded to a folded diaphragm. By means of the top-to-bottom folds of the diaphragm, the current flows vertically in and out of the page, as represented here.

For the no-signal condition, the folded diaphragm is as shown in Figure 16-1b. For positive current flow the diaphragm takes the position shown in Figure 16-1c, which produces positive pressure to the left. For negative current flow the diaphragm folds take the position shown in Figure 16-1d. (In the convention shown here, the dot indicates current flowing out of the illustration, and the cross represents current flowing into the illustration.)

The notion of air motion transformation is an accurate one in that the air particle velocity outward from the folds is several times that of the folds themselves.

The driver is normally used in the frequency range above 1 kHz. The efficiency of the driver is moderate, and the output can be raised to significant levels by placing a horn on one side of the diaphragm. A wide-range system based on the AMT topology was developed in the 1970s by the ESS Corporation.

16.2.2 Traveling Wave Radiators

In most dynamic transducers, pistonic action is considered ideal. Traveling waves in the diaphragm are generally regarded as detrimental, since the result is usually response irregularities. However, advantage may be taken of traveling waves if they can effectively be damped at the fixed termination of the diaphragm.

Figure 16-2 shows a section view of a transducer developed by Lincoln Walsh. The voice coil is attached to the apex of a deep cone which has been positioned vertically and is free on all sides. High-frequency radiation parallel to the cone's motion is effectively absorbed in the lining of the enclosure and in the cone's surround, and only the motion of the cone at right angles to the up-down motion will produce useful acoustical output at high frequencies.

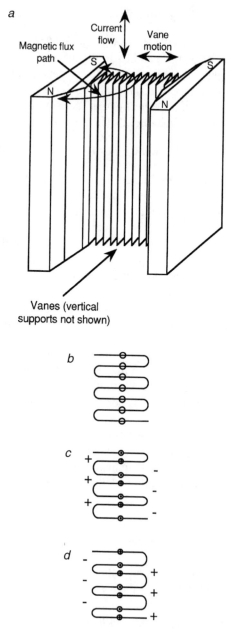

Figure 16-1. The Heil air motion transformer. Perspective view (*a*); Section view, no-signal condition (*b*); section view, positive signal condition (*c*); section view, negative signal condition (*d*).

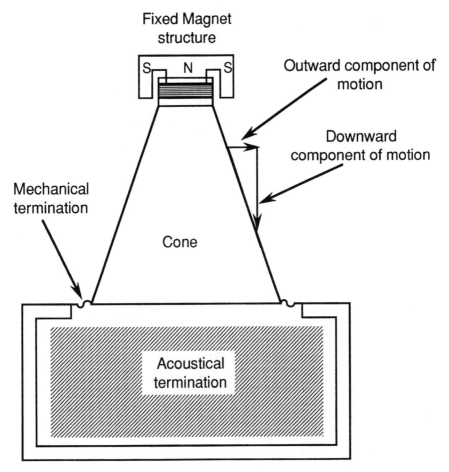

Figure 16-2.   The Walsh radiator, shown in section view.

At low frequencies the cone will produce essentially omnidirectional radiation. At progressively higher frequencies, traveling wave motion in the cone will result in a gradual shift in radiation outward from the cone, and ultimately, at the highest frequencies, the radiation will be solely from a very small portion of the cone adjacent to the voice coil. The device thus operates as a full-range radiator, as the voice coil effectively becomes decoupled from the mass of the cone.

There is nothing trivial about the design of the moving system, and it normally consists of several distinct sections, each optimized for a specific portion of the frequency spectrum. When well executed the Walsh driver can produce excellent response with moderate efficiency.

Another approach to traveling waves is shown in front view in Figure 16-3*a*. This design consists of a voice coil attached to a large area of plastic under

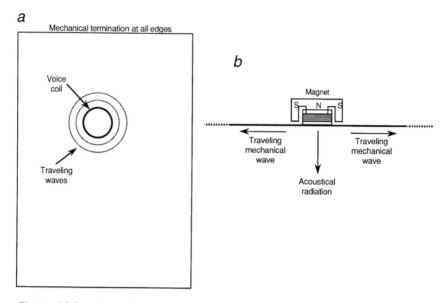

Figure 16-3. A traveling wave radiator. Front view (*a*); horizontal section view (*b*).

tension which has a damped termination at its edges. Motion of the voice coil is translated into outwardly radiating rings (Figure 16-3*b*). High-frequency signals are attenuated fairly quickly due to the mass of the diaphragm and are thus radiated from a small area around the voice coil. Successively longer wavelengths are radiated from larger portions of the diaphragm, and the dipole pattern of the system is consistent over a wide frequency range. The system is of moderate efficiency, and the chief problem in manufacture is maintaining consistent and uniform tension in the diaphragm.

16.2.3 Systems Using Compound Radiating Surfaces

The three transducers discussed here exhibit complex motion with more than one degree of freedom and as such exhibit radiation patterns that are unlike the ideal single degree of freedom radiators that acoustical engineers hold in esteem. While these devices may fall short of ideal pulsating spherical or cylindrical response, they can exhibit more uniform pattern control than typical cones and domes over fairly large portions of the frequency range.

The Linaeum driver is shown in front view in Figure 16-4*a*. The half-cylinder sections are fixed at their outside edges and driven along the other (middle) edges. The nature of the compound motion of the cylinder is shown in Figure 16-4*b*. It is important in this design, as in all others with fixed terminations, that reflections from the termination back toward the driving point be properly damped.

Allison (1995) describes a HF dome radiator with an inverted outer section

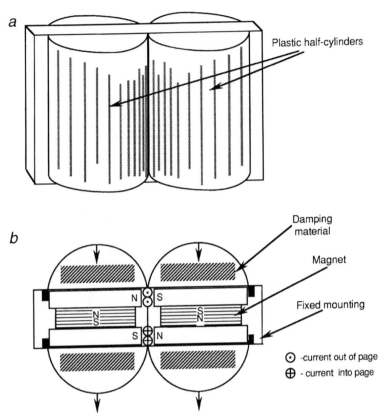

Figure 16-4. The Linaeum driver. Perspective view (*a*); horizontal section view (*b*).

that exhibits radial components of movement, and as such does not exhibit the typical narrowing of HF pattern control of a single-degree-of-freedom dome. A section view of the driver is shown in Figure 16-5*a*, and the normalized off-axis HF response is shown in Figure 16-5*b*. The hoop stresses associated with warped surface motions (see Chapter 4) are minimized, inasmuch as the displacement of the diaphragm at high frequencies is extremely small. For comparison purposes the normalized off-axis response of a conventional 25-mm (1-in.) HF driver is shown in Figure 16-5*c*.

The complex radiator shown in Figure 16-6 is made by the MBL Company of Germany. Shaped like a seamed football set on end, it is fixed at one end and driven at the other along its vertical axis. The seams are flexible and allow the structure to pulsate in and out when excited by the voice coil. Drivers of this type are limited in their frequency range, and a three-way wide-band system has been commercialized. Obviously, the radiation pattern is uniform in the horizontal plane.

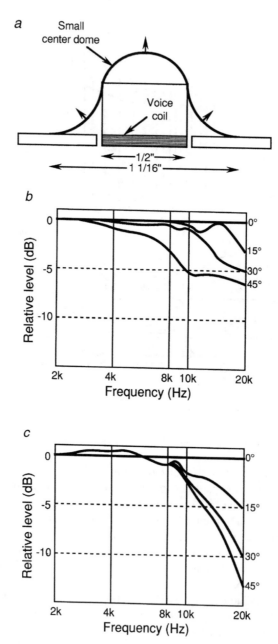

Figure 16-5. Section view of the Allison HF radiator (*a*); normalized off-axis response curves (*b*); normalized off-axis response curves for a typical 25-mm dome (*c*). (Data after Allison, 1995.)

Fixed
mounting

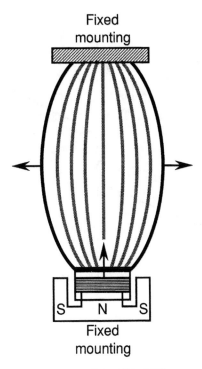

Fixed
mounting

Figure 16-6. Section view of the MBL 360° compound radiator.

16.3 Piezoelectric and Related Devices

Certain crystal materials will bend, twist, or shear slightly if electrical signals are applied to opposite surfaces on them. In general, displacement is proportional to the applied voltage. The principle has been used for decades in low-cost microphones, mechanical sensors, and HF loudspeakers. Details of a piezoelectric HF driver are shown in section view in Figure 16-7. Here, a small bending element is attached to a stiff diaphragm that resonates in the HF range. The driving force is the result of the effective mass of the bending element times the acceleration imparted by the driving signal. The range of flat power output may extend approximately one octave above the diaphragm resonance frequency, although the on-axis output can be extended to a somewhat higher frequency. These devices are not very smooth in their response, but this is a matter of secondary concern in many applications. When used in multiples, and with horn loading, the output capability is fairly high.

The high polymer material (HPM) HF unit is shown in Figure 16-8. As developed by Pioneer (Tamura, 1975), this HF transducer consists of a sheet of polyvinylidene fluoride curved into a cylinder. The signal is applied to the two

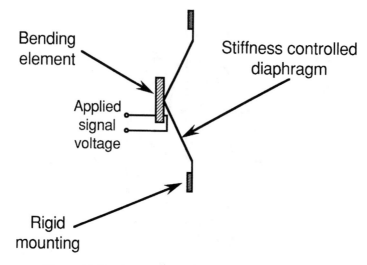

Figure 16-7. Section view of a piezoelectric HF radiator.

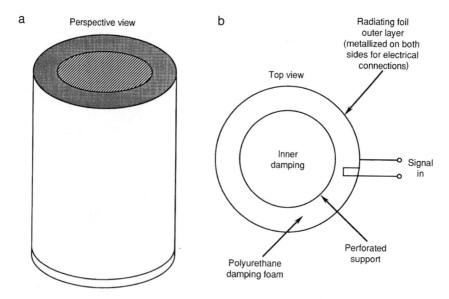

Figure 16-8. The Pioneer HPM driver. Perspective view (*a*); top view (*b*).

sides of the material and the plastic effectively stretches and contracts under the electrical excitation, producing 360° radiation. The device is very uniform in response from 2 to 20 kHz.

16.4 Ionized Air Devices

A section view of the Klein (1952) Ionovac is shown in Figure 16-9*a*. In this complex device, a high frequency-high voltage signal is fed to inner and outer electrodes separated by a quartz cell. The ionized air in the cell glows with a blue light as a result of what is called corona discharge, and this feature alone is sufficient to entice anyone attracted to the exotic. If the HF signal is amplitude modulated, then the volume of ionized air will vary accordingly, creating instantaneous pressure variations. The acoustical output is relatively low, and a horn is normally used to increase the output. The curves shown in Figure 16-9*b* were made some years ago on the Ionovac HF system manufactured by the Du Kane Corporation. While fairly uniform above 3 kHz, the maximum output capability was limited at high frequencies, exhibiting noticeable signal compression.

Over time, the quartz cell and inner electrode will erode and must be replaced; and in general the complexity of the system far outweighs its advantages. However, during the decades of the fifties and sixties, the Ionovac represented the leading edge in HF performance.

The so-called corona wind loudspeaker (Shirley, 1957) operates with high-voltage dc bias across which is impressed a high-voltage audio signal. When the positive electrode is sharply pointed, and the negative electrode blunt, as shown in Figure 16-10*a*, there will be a small but steady flow of air from positive to

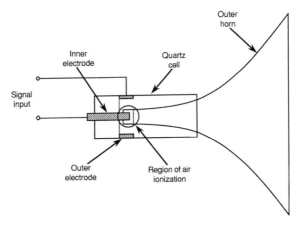

Figure 16-9. Design details of the Du Kane Ionovac HF system (*a*). Response curves of the Ionovac system measured at 1.2 m; scale at left indicates the signal input to the system's modulation transformer (*b*).

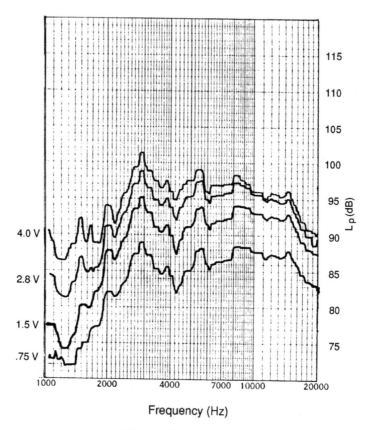

Figure 16-9. *Continued*

negative, hence the term "corona wind." The air flow can be modulated by an audio signal, as shown.

If both electrodes are sharply pointed, there will be no wind under no-signal conditions. The corona action requires a dc polarizing voltage of 10–20 kV, while the audio signal usually varies over a range from 1 to 2 kV.

A typical system will be made up of many individual cells located on a plane. Typical response of a prototype system is shown in Figure 16-10b. The dip in response at 8 kHz occurs when the spacing between the pointed electrodes is equal to $\lambda/2$. The dc operation of the form shown in Figure 16-10a has been used in the manufacture of air filtration devices.

A word of caution: the various ionized air systems generate ozone and as such may constitute an environmental hazard.

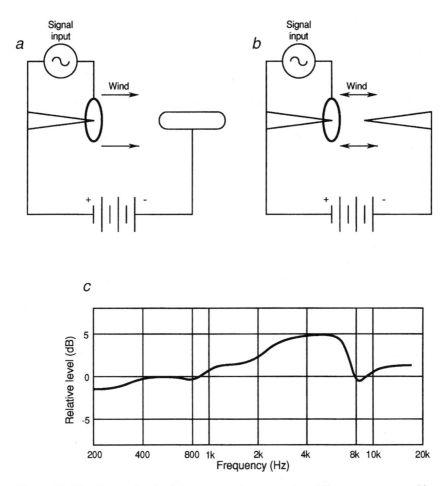

Figure 16-10. Design details of the corona wind system (*a* and *b*); response curves (*c*).

16.5 The Air Modulator

The air modulator is a very practical device, but not for general home high-fidelity use. Essentially, it is a means of using a relatively low-power dynamically driven valve to control the flow of a relatively high power source of compressed air (Hilliard 1965). One of the first embodiments of this principle was the Auxeto-phone, which was developed by Parsons during the acoustical phonograph era. This device used mechanical power from acoustical recordings to drive an air valve for greater acoustical output. Stability was a problem with this early pneumatic system; particles of dust could get into the delicate valve, resulting in hisses and other noises.

One form of the modern-day version the air modulator is shown in Figure 16-11. An inner movable sleeve with openings fits smoothly inside an outer fixed casing. In the rest position, there is a portion of overlap between the two sets of openings, and a steady stream of air is emitted through the system. When the inner sleeve is moved back and forth by a dynamic assembly, or other linear actuator, there will be an instantaneous change in the volume velocity of air, and hence a change in pressure. A major design challenge in such systems is to shape the various openings to minimize turbulence and air noise, while optimizing the system's linearity over the desired output range.

The air modulator is routinely used for high-pressure acoustical output of noise signals useful for environmental testing, where tens of thousands of acoustical watts are necessary to simulate the LF sound intensities developed by modern aircraft and space rocketry. In these applications the air overload can be excessive, as pressures are developed in the range of 150–160 dB and beyond. In such applications, system linearity is not a primary concern; rather, a particular noise power spectrum at some target level may be the only concern.

There are distinct opportunities for the air modulator in producing very-high-pressure, VLF acoustical signals in special venues such as motion picture theaters, rock concerts, and theme parks. In these cases the cost of maintaining a continuous source of compressed air is negligible, and a value analysis of the system, as compared to standard technology, might favor its use. Much work remains to be done in this area.

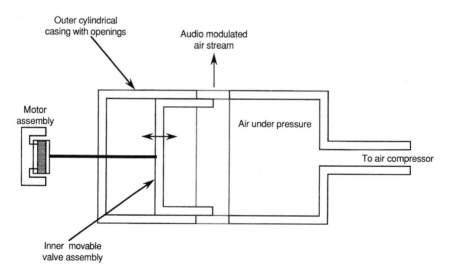

Figure 16-11. Section view of an air modulator.

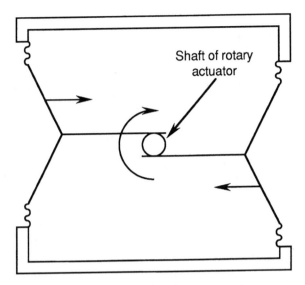

Shaft of rotary
actuator

Figure 16-12. Section view of a loudspeaker system using a rotary actuator.

16.6 Rotary Actuators

The rotary actuator is extensively used in mechanical indexing and control systems, and its application to acoustical systems follows as a natural consequence of this. The basic approach is shown in Figure 16-12, in which the the rotary actuator converts normal alternating signals into corresponding rotary ones. The rotary action is mechanically converted into linear action required to move cone radiators. Depending on the mechanical transformation ratio between angular and linear motion, fairly large excursions can be achieved, limited only by the mechanical suspensions used in restraining the radiating cones. The bandwidth limitations of the method normally restrict its application to LF operation.

16.7 Digital Loudspeakers

In concept, a digital loudspeaker converts a parallel digital signal directly into pressure output pulses, each bit physically scaled according to its position from least to most significant bit. For a 16-bit system, the required scaling range is 65,536 to 1. Covering this range with concentric or otherwise acoustically aligned radiating elements hardly seems possible or practical. However, a lesser-bit approach might work well enough to enable a hybrid system to be designed.

The method shown in Figure 16-13 shows an array of four electrostatic elements with areas scaled by powers of 2, each driven by a parallel digital signal. The least significant bit drives the center element, while the outermost element is driven by the most significant bit. The composite element can be used as a

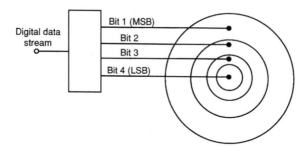

Figure 16-13. View of a 4-bit electrostatic digital transducer (LSB = least significant bit; MSB = most significant bit.)

headset driver, with the acoustical output summation taking place directly at the listener's ear. This device works for some types of headphone communication, where the high noise level associated with it may be tolerated.

An additional approach is shown in Figure 16-14. The similarities with the previous example are evident, and the same problems of scaling and summing of the individual acoustical output bits remain. In this model, the direct digital-to-analog conversion is used only for the 8 most significant bits; the lower-value bits are converted in the normal manner and fed to the ninth coil. This hybrid system thus is acoustically digital only for the upper 48 dB of its output level range.

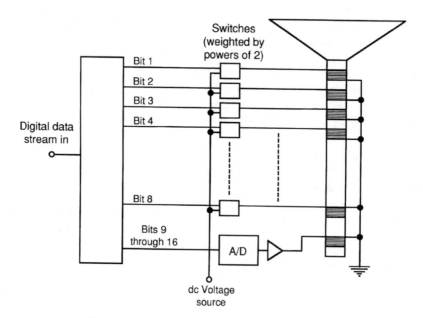

Figure 16-14. View of a hybrid digital loudspeaker.

Bibliography

Allison, R., "Imaging and Loudspeaker Directivity: To Beam or Not to Beam," Preprint No. 4095; presented at the 99th Audio Engineering Society Convention, New York, October 6–9, 1995.

Augspurger, G., "Theory, Ingenuity, and Wishful Wizardry in Loudspeaker Design—A Half-Century of Progress?" *J. Acoustical Society of America*, Vol. 77, No. 4 (1985).

Bost, J., "A New Type of Tweeter Horn Employing a Piezoelectric Driver." *J. Audio Engineering Society*, Vol. 23, No. 10 (1975).

Collums, M., *High Performance Loudspeakers*, Wiley, New York (1991).

Hilliard, J., "High-Power, Low-Frequency Loudspeakers," *J. Audio Engineering Society*, Vol. 13, No. 3 (1965).

Klein, S., *L'Onde Electrique*, Vol. 32 (1952), p. 314.

Shirley, G., "The Corona Wind Loudspeaker," *J. Audio Engineering Society*, Vol. 5, No. 1 (1957).

Tamura, M., et al., "Electrostatic Transducers with Piezoelectric High Polymer Films," *J. Audio Engineering Society*, Vol. 23, No. 1 (1975).

Index